SEVER THE SKY

EVOLUTION OF SEVERSKY AIRCRAFT

by edward t. maloney

Library of Congress Card Number 79-23714   I S B N Number 0-915464-01-2

Printed in The United States Of America

PLANES OF FAME PUBLISHERS

WORLD WAR II PUBLICATIONS
Post Office Drawer 278
CORONA DEL MAR,CALIF. 92625

CONTENTS

DEDICATION

This book is humbly dedicated twofold: to Maj. Alexander P. de Seversky, a very gallant public servant of civil and military aviation designs. His strong will and foresight gave America some of its most outstanding aircraft pursuit designs. Seversky aircraft set many coast to coast speed records and his name became a legend in his own time; also to the men and personnel of the 17th and 34th Pursuit Squadrons, 24th Pursuit Group who flew and fought with the Seversky P-35A in the Philippines 1941-1942. This courageous band of fighting men flew against overwhelming odds and a superior enemy, but prevailed during these extreme hardships as Americans have always done when the chips are down. May their memory never be forgotten.

COVER: Last of the flying Seversky's. Model 2PA-204A (AT-12) over Planes of Fame Museum, Chino Airport, California.

INTRODUCTION

This book is an attempt to provide the aviation enthusiast, historian, and model builder a pictorial review of many outstanding aircraft designs of one of America's premiere aircraft companies of the 1930's. The products of this company made great contributions to American Pursuit Aviation in a period long remembered by those who took part but are seldom recalled today by new generations.

The aircraft of this period are lost now and only four examples remain in museums for you and I to view. The time period in which these aircraft fought the four winds was a colorful and fascinating period, and one which will never be seen again, but come with me and lets go back and recall some of the outstanding aircraft from this period.

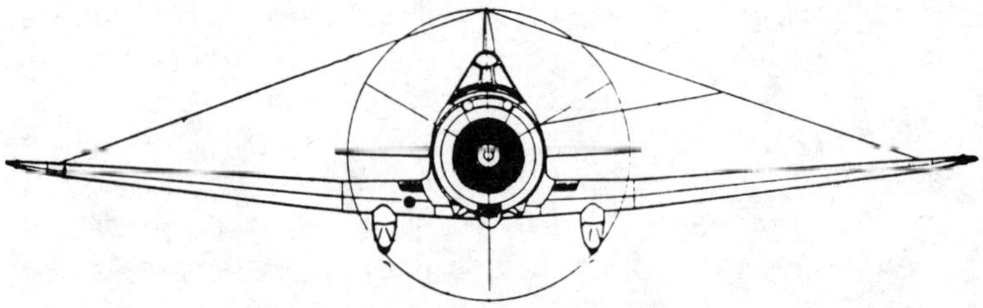

IN APPRECIATION

A special vote of thanks goes to the Air Force Museum, Mr. Charles Worman and his wonderful staff of co-workers, Roger Besecker, Stephen Hudek, Dusty Carter, Robert Hirsch, and Paul Matt who all gave of their time and valuable assistance. Thanks also to the following persons and companies who graciously provided information, photographs, and details.

Republic Aircraft Corp.
National Air & Space Museum
U.S. Air Force
N.A.C.A. (N.A.S.A.)
Paul Matt & Aero Album
Don Lykins
Lamar Gillette
Jay Miller
James Copeland
Tom Piedmonte
George Armstrong
Peter Bowers
Mike de Marino
Leo J. Kohn
Warren Bodie
Robert Mikesh
Ken Rust
Stephen Kanyusic
Anthony Yusken
Gene Krope
Leonard Godbold
Fred Quick
Bill Larkins
Robert Esposito
Gen. John Dolny U.S.A.F.
Don Thorpe
Jim McDonnell
Frank Strnad

MAJOR ALEXANDER P. DE SEVERSKY

He was born in Tiffis, Russia, June 7, 1894, and christened Alexander Nicholaivitch Pro cofieff de Seversky. He graduated from the Russian Naval Academy in 1914 and was later selected for duty as a military aviator. During the following three years he became an "Ace" with thirteen victories, but lost a leg in action. Fitted with an artificial limb he continued to do battle with the enemy. He became the leading Naval Ace and received all the honors that could be bestowed on him.

In 1918, he was sent as a member of the Russian Naval Mission to the United States. When Russia dropped out of World War I, he offered his services to the United States Government and was appointed aeronautical engineer and test pilot. In 1921 he acted as advisor to Gen. William "Billy" Mitchell during the demonstration of an aircraft sinking a battleship. He was later appointed consulting engineer to the War Department by the Secretary of War.

Maj. Seversky was proud of his SEV-3 Amphibian. Shown here with canopy hatches open. Aircraft accomodated three persons in spaceous interior.

In 1927 he became an American citizen and was later commissioned a Major in the U.S. Air Corps Reserve. Seversky is credited with designing the world's first fully automatic bombsight.

In February 1931, he founded the Seversky Aircraft Corp., with Wall Street backing. He was founder, president, designer, and chief test pilot. This organization later became Republic Aviation Corp., which today is a division of Fairchild. He was President and Director for seven years and during that time set many speed records. Under his leadership his company built some of America's best known pursuit planes. He designed, built, and flew the fastest amphibian aircraft in the world, the first low wing monoplane basic trainer for the Army Air Corps, and pioneered the development of the first turbo supercharged, air cooled engine fighter. His advanced ideas contributed greatly to the defense of this country. His pleas for more planes, pilots, bigger and better pursuit aircraft, and faster, longer range aircraft were finally heard.

In December 1940 he was presented the coveted Harmon Trophy by President Franklin D. Roosevelt for his outstanding contributions and achievements in the field of aviation. Major Seversky was recognized as one of our greatest experts on strategy and tactics of air warfare.

SEV-3 was later modified with 3-bladed propeller for better performance.

Seversky was also a genius at acquiring good talent, and with Alexander Kartveli as his Chief Engineer, designed some of America's most modern pursuit aircraft. They combined their knowledge to develop the first turbine driven super-charger in combination with an air cooled engine. This plane was designated XP-43 "Lancer." It received a quantity order by the Air Force and was used in combat by the Chinese Air Force during World War II.

At a meeting of the Board of Directors in 1939 Major Seversky was voted out of office by the Board. Several members of the Board charged that too much time and money was going into experimental work which resulted in fewer quantity orders. This was proving too costly for the firm as it had been in the red for several years.

Seversky defended his position by pointing out that only by experimenting can great results be achieved. He was outvoted by the Board and forced to resign his presidency. The company was reorganized as Republic Aviation Corporation. Ralph Damon was elected President and Alexander Kartveli was retained as Vice President and Chief Engineer.

Major Seversky was bitterly disappointed. However, he devoted his time to writing his famous book "Victory Through Air Power." Major Seversky's book was a best seller. It was a large chronicle of War Department blunders and military shortsightedness, but its theme was not despair. Its message to the American people was one of soaring hope and strength by giving America "Victory Through Air Power." Seversky was not afraid of speaking out if he felt right was on his side. He toured the country speaking and lecturing. Throughout the land his was a voice to be heard and heeded. Walt Disney later made a film on Seversky's book "Victory Through Air Power."

Maj.Seversky by SEV-3M of 1935. Note canopy has been enlarged, smaller float rudders, and ADF loop added. Ship was Maj. Severskys pride and joy.

SEV-3XAR was original SEV-3 amphibian modified to land plane.

During World War II Major Seversky was awarded the "Medal of Merit," the highest wartime honor for a civilian. He won this for his great contribution in developing the long range escort fighter. Later President Harry Truman awarded him a citation for his vigorous support of the Air Force.

Major Seversky held many patents and copyrights. Working with Elmer Sperry, Maj. Seversky laid the foundation for all gyroscopically stabilized flight instruments. This made the automatic pilot possible.

After World War II Major Seversky represented Secretary of War, Robert P. Patterson at the first peace time Bikini atom bomb test.

As years went by, Maj. Seversky turned his talents toward ecological matters. He invented the Seversky wet electrostatic precipitator to extract impurities from the air and from industrial emissions. He still remained active giving lectures and talks to various aeronautical organizations and was president and chairman of the board of Seversky Electronatom Corp., the manufacturer of his ecological device. Seversky was an advisor to the late Aristotle Onassis. He continued to be active almost until the day he died at the age of 80 years in New York City, on August 25, 1974.

He had been married to the late Evelyn Olliphant of New York.

Seversky 3XAR competed in Air Corps trainer competition.

The noted aircraft designer Alexander Kartveli

ALEXANDER KARTVELI

Alexander Kartveli was a fellow countryman of Major Seversky. He fought in World War I as an artillery officer of the Imperial Russian Army. After the war he studied aeronautical engineering in Paris, France.

In 1928, he came to the United States and became a citizen shortly thereafter. Kartveli held a Doctor of Science Degree in Aeronautical Engineering. He was Major Seversky's design assistant on many of the Seversky aircraft designs built at Long Island, New York.

Following the company re-organization, Kartveli was made a Vice President and given charge of the design bureau as chief engineer. He was given a free hand to continue the development of fighter aircraft. After a fighter design conference at Wright Field, Kartveli dropped the new XP-44 "Warrior" design. A million dollars had been spent on design and detail engineering drawings, research, and mock-up experiments.

Kartveli returned to his drafting board and came up with the XP-47B design, known to all as the "Thunderbolt." The outstanding success of this world famous fighter during World War II gave the recognition to Alexander Kartveli he justly deserves. Alexander Kartveli passed away in the summer of 1974 just a few months prior to Alexander P. de Seversky.

Seversky won world wide acclaim for his SEV-3 Amphibian design.

I THE BEGINNING

Major Alexander P. de Seversky formed the Seversky Aircraft Corp. in February 1931 at Long Island, New York, with Alexander Kartveli as chief engineer. Having no facilities to start manufacturing, he contracted out his first aircraft to Edo Aircraft Float plant at College Point, Long Island. The first designed plane was built there and was designated SEV-3 Amphibian, Registration No. X-2106. The SEV-3 was a beautifully designed amphibian and it made its first public appearance in June 1933. From a humble beginning in a former Long Island potato patch, the SEV-3 became a record breaking aircraft.

The SEV-3 was three place, all metal, low wing amphibian. It was powered by a Wright J-6 engine of 420 horsepower and did 190 m.p.h. at sea level. It had twin floats with retractable wheels within the main floats. The wheels were recessed into slots in the floats, and were directly supported from the wings independent of the floats. They were extended from the floats by hydraulic power. When the SEV-3 was operating from water, the wheels were retracted. This operation automatically locked the floats in position. When operating from land fields with the wheels fully extended the floats became free to move, and when a normal landing was made, small auxiliary wheels attached to the rear float water rudders pushed the floats up until the aircraft was running in a normal fashion on the main wheels and tailwheel.

On October 9, 1933 Maj. Seversky set an international speed record in this aircraft of 179.7 m.p.h.

In 1934, the SEV-3 was converted to a landplane with the removal of the twin floats and installation of a spatted fixed landing gear. It was modified to SEV-3XAR, an experimental aircraft for the Army. The complete canopy was raised in height, and a new Wright R-975 engine of 350 horsepower was installed. The engine was moved further forward for better weight and balance with the floats removed. This aircraft was demonstrated for Army Air Corps basic trainer competition at Wright Field. It won the competition and brought the Seversky Corp. its first large Army contract order for 30 Seversky BT-8's.

Maj.Seversky taking off SEV-3 amphibian on its 1st test flight at Jamaica Bay

Later, the aircraft again was modified back to an amphibian and redesignated SEV-3M. A new Wright R-1820 engine of 750 h.p. was installed. It was in this aircraft that Maj. Seversky set a new international speed record for amphibian aircraft of 230.4 m.p.h. on September 15, 1935. This record breaking speed demonstration resulted in several good sales for the Seversky Company to foreign governments. This prototype model was later sold to a Mexican Army officer in 1938. It is believed it was again sold and sent to Spain to fight on the Republican side during the Spanish Civil War.

The second Seversky aircraft was the SEV-3M Registration No. X-18Y. The "3M" designation stood for three place military. It is doubtful if it ever operated in this configuration. It became the SEV-2XP, an experimental two place Pursuit, with a spatted fixed landing gear and a Wright R-1820 engine of 750 h.p. It was hurriedly finished in order to participate in the 1935 Wright Field Pursuit competition, but due to an accident enroute to Wright Field, it never reached its destination. The Design competition was postponed which made the time needed for the Seversky Company to rebuild the aircraft. When it was returned to the factory, the aircraft was modified into a single place pursuit, with a retractable landing gear and Wright R-1820 engine. Horsepower was increased to 850. The aircraft was redesignated SEV-1XP.

It was test flown successfully, and appeared in August 1935 at Wright Field Pursuit Competition, but no Army contract was received. The aircraft returned to the factory where a new Pratt-Whitney R-1830 engine of 850 h.p. was installed. In April 1936, it again appeared in the pursuit competition at Wright Field, and this time it was awarded a contract by the Army under the military designation P-35. The Prototype model X-18Y was flown by Maj. Seversky and Jacqueline Cochran. They set several speed records between some of the major cities in the United States. This aircraft was also flown by Frank Sinclair in both the 1937 Bendix and Thompson Trophy Races. In 1939 it appeared in the United States aviation exhibit at the New York World's Fair.

Seversky 1XP featured retractable landing gear and Wright R-1820 engine.

2nd Seversky aircraft,the SEV-2XP was 2 place pursuit for June 1935 competition. Accident prevented its appearence. It returned to factory and became SEV-1XP.

Seversky 1XP just prior to being displayed at New York World's Fair. Thousands of viewers were permitted to see this famous record breaking aircraft however Army Air Corps had Maj. Seversky paint all canopy glass white so no interior photographs could be made.

Seversky 1XP originally mounted Wright R-1820 engine but was later modified to P&W R-1830-9 engine of 850 h.p. Aircraft became Maj.Severskys favorite.

Maj.Seversky flying his high performance SEV-1XP above New York skyline

Seversky 1XP won April 1936 Army Air Corps design competition. Order was placed by Army for 77 P-35 pursuit aircraft.

Maj. Seversky stands by 1st of three SEV-3M-WW's built for Columbia

In August 1935, Maj. Seversky received his first foreign contract for three of his Seversky SEV-3M-WW Amphibians from Columbia. Still not having an aircraft factory in which to build these aircraft he signed a contract with Kirkham Aircraft Company to build them. Months passed and it soon became very evident that Kirkham was not going to be able to finish the three amphibian aircraft. Maj. Seversky took over an idle building adjoining the Kirkham plant and worked to complete the three aircraft as ordered. He delivered them to Columbia on time as per contract.

During the early Seversky years both Grumman Aircraft Corporation, and Seversky Aircraft Corporation were building aircraft as neighbors; they shared the same airfield, the same paint shop building, and other joint facilities, at Farmingdale, Long Island.

A great rivalry grew up between the employees of the two companies, both of whom were intensely loyal and proud of their respective organizations.

Both companies were tenants of American Aircraft and Engine Companies' property at the time when a dispute over options to buy the property arose. Grumman pulled up stakes and moved away to Bethpage, Long Island, New York in 1937. As soon as Grumman had left, the ever growing Seversky company moved in and occupied the vacated building and other facilities, thus occupying all the remaining property except a small portion occupied by Ranger Aircraft Engine Company.

Maj. Seversky was busy demonstrating his products the world over but his star began to fade. Leaving business management in the hands of others his company went downhill financially. Wrong decisions were made and grave deficiencies in management and conduct of the plant brought about severe criticism. A bad relationship with the notorious Miranda-Zelcer combine, and sale of 20 Seversky aircraft to Japan put the entire Seversky Company in a very bad light with the American Government. Maj. Seversky was a salesman and he reasoned if the Army Air Corps would not buy his new designs, he would sell to overseas buyers. Business was business.

In November 1938, Wallace Kellett of Philadelphia, Pa. and former president of Kellett Autogiro Company, was made executive vice president under Maj. Seversky, the founder and President. A swift reorganization followed, including reducing swollen payrolls, obtaining new bank financing, and getting new business contracts.

Three SEV-3M-WW were delivered to Columbia in 1935

Seversky 3M-WW was 1st export order received from Columbia

Seversky 3M-WW ship #3 in Columbian colors,awaits delivery.

SEVERSKY P-35 PROTOTYPE

Up until 1934 all production Army Air Corps Pursuits were of biplane design. With the introduction of the Boeing P-26 "Peashooter" the way of the future was the all metal, low wing aircraft with fully enclosed cockpit and retracting landing gear.

In 1935, the Army Air Corps Material Division began to search for a replacement for the P-26, then in service. A Wright Field Pursuit competition was announced and bids were invited. When the bid opening was held in May 1935, only the Curtiss Company had submitted a design proposal. A second competition was announced and Maj. Seversky said he would send an entry for this August 1935 competition. Seversky entered his new SEV-2XP design. The SEV-2XP was experimental, two place Pursuit, with a spatted fixed landing gear and Wright R1820 engine 750 h.p. This aircraft was damaged enroute to the competition. Two other entries came into the competition, Consolidated and Vought; each submitted a prototype design to compete in this competition.

Chief competitor to P-35 was Curtiss Model 75 which served as prototype P-36.

The Curtiss Hawk Model 75 prototype Reg. No. X-17Y design was initiated in late 1934. It orginally featured a Wright R-1670 engine of 900 h.p. Seversky entered his aircraft, the 2-XP Reg. No. X-18Y. It arrived at Wright Field June 18, 1935. It was returned to the factory and rebuilt into a single seat pursuit with retractable landing gear. A newer Wright R-1820 engine of 850 h.p. was installed. Redesignated SEV-1XP it arrived back at Wright Field on August 15, 1935. It was favored by Army Air Corps and would have received a contract, but Curtiss Company officials protested that an unfair advantage had been obtained by postponing the design competition in Seversky's favor. This resulted in a third competition and the date was announced April 15, 1936. When the final bids were opened the Curtiss Company was the low bidder: $29,412 for P-36 aircraft in lots of 25 planes, and $14,150 each in lots of 200 planes. The Vought Company submitted a bid of $34,148 each in lots of 25 planes and $16,051 in lots of 200 planes for its Model V-141 design. This design was orginally a Northrop Army Model XP-948 or Northrop 3A. This prototype took off from the Los Angeles factory, headed out over the Pacific and was never seen again. The design rights were then purchased by Vought and the model was designated V-141. It mounted a Pratt-Whitney R-1535 engine of 750 h.p. It had a beautiful flush mounted retractable landing gear and was the smallest entry in the competition. However the V-141 had a short coupled fuselage and suffered tail vibrations. It was sold to Japan in 1937. The Japanese Mitsubishi Zero Fighter designer, Mr. Jiro Hirokoshi has stated that the system of landing gear retraction was based on the Vought V-143 design but that this was the only contribution to his design. There were many radial engine low wing all metal fighters built during these pre-war years. Each one was built in response to a particular need. The Zero Fighter went on to achieve aerial supremacy and became Japan's best known fighter of the Pacific War. It would later meet the Seversky P-35A in the skies above Clark and Iba Fields, Philippines, but that is getting ahead of our story.

Consolidated entered a single place version of its PB-2A design but it was disqualified due to its destruction in a test flight. It was the only liquid cooled engine entry in the competition.

Seversky bid $34,900 per aircraft in lots of 25 airplanes, and $15,800 in lots of 200 airplanes.

The Material Division of the Army Air Corps asked both Seversky and Curtiss for alternate performance tests with the new Pratt-Whitney R-1830 Wasp engine

Vought V-141 competed against P-35 at Wright Field. Ship was later sold to Japan. Max. speed was 292 m.p.h. at 11,475 feet.Range was 800 miles.

whose smaller diameter provided less drag and improved visibility. Maintenance problems with the Wright R-1820 had plagued both Seversky and Curtiss aircraft during the competition. The Curtiss Model 75 design had four engine changes during the design competition trials and only achieved 285 m.p.h. of the 295 m.p.h. guarenteed by Curtiss. The Seversky 1XP entry hit 289 m.p.h. instead of the 300 m.p.h. promised. When Seversky changed to R-1830 only 277 m.p.h. was achieved, for the R-1830 developed 740 h.p. instead of the 850 h.p. guarenteed by the manufacturer. Nevertheless, the Air Corps specified the R-1830 for both aircraft when production orders were placed.

The Air Corps did not abandon the Wright R-1820, rather it desired a second source of supply. The Wright Aeronautical Company was building engines for the B-17, B-18, and several other aircraft. This proved to be a wise decision, for when war came, there were two reliable radial engines to meet the emergency in place of one single source.

Officially, there never was an XP-35. The Seversky 1XP served as the prototype test model and as Seversky's bid was accepted, the production model became the P-35. The Air Corps on June 16, 1936 ordered 77 P-35's from Seversky as the winner of the design competition. A year later they ordered 210 P-36's from Curtiss as war clouds gathered over Europe.

The production P-35's had P&W R-1830-9 Wasp engines. It differed from the prototype by having modified wheel fairings, and the oil cooler intake was moved from the left side to the bottom of the engine cowl. The 1st production P-35 S/N 36-354 had fully enclosed wheels and no wing dihedral but later production models had open wheel fairings and dihedral.

The rearward retracting landing gear of the P-35 was only partially enclosed within the wing when fully retracted. This resulted in two streamlined bulgeous fairings below the wing. These protuberances actually increased performance and this phenomenon is known as a Whitcomb Body. If a wheels up landing had to be made due to some malfunction, there was less damage done than in other aircraft as the landing gear was already partially exposed and the wheels would rotate freely with ground contact.

The P-35 was easy to fly and was an eyecatcher wherever shown. Its 950 h.p. P&W R-1830-9 gave it a good rate of climb and a top speed of 281 m.p.h. which was almost 50 m.p.h. faster than the P-26A "Peashooter." Its range of 1000 miles was more than adequate to fly in defense of the shores of the United States. Performance was further improved by the introduction of 100 octane fuel by Shell Oil Company. By European standards the P-35 was lightly armed having only one .30 and one .50 caliber Browning machine gun mounted in the cowl above the engine and firing through the propeller. The last P-35 was delivered to the Air Corps in August 1938.

The P-35 was a soundly built aircraft structure. Ease of maintenance was designed into the aircraft. Most areas of the aircraft were readily available and the aircraft was spacious and roomy. A mechanic could be taken along in flight as a passenger by fitting him in the baggage compartment where he could sit upright.

The early P&W R-1830-9 powerplant was a reliable engine, but like most radial engines of that period, it leaked oil. After long flights the aircraft had to be completely washed down with solvent, for if it was allowed to stand overnight, it only made matters worse for cleaning the following day.

Production P-35's were initially equipped with shotgun type engine starters.

Northrop 3A took off on its 1st flight from Mines Field July 10,1935 and disappeared over Pacific Ocean. Design was later sold to Vought Aircraft.

Maj. Seversky was one of the ablest of flyers and aeronautical designers

These starters cracked engine gears, frequently jammed and became a real problem. They were replaced by new all electric starters.

The P-35 had what is known as a "wet wing," literally the entire wing was sealed up and made into a large gas tank. This accounted for the P-35's remarkable long range, but after several months in service the tank sealant would dry up and fuel leaks would appear from the belly wing plates and screw holes. Despite these faults the P-35 continued on in service and gave a good account of itself with the Air Corps.

Army Air Corps ordered 77 Seversky P-35's in 1937

SEVERSKY BT-8 "BASIC TRAINER"

The BT-8 was built under an Army contract in 1936 at the Farmingdale, Long Island plant. The all metal BT-8 was designed to be easily built on a modern production line basis. In 1935 Seversky had won the Basic Trainer competition at Wright Field and an order for 30 BT-8 trainers plus spares was placed. The BT-8 was the first modern, low wing trainer produced for service with the Army Air Corps.

BT-8 was powered by P&W R-985 engine. Top speed was 175 m.p.h.

The BT-8 had a Pratt-Whitney R-985-11 engine of 400 h.p. The number one production BT-8 was delivered to Wright Field in February 1936. The second production model was delivered April 1936 and assigned to the Air Material Division of the Air Corps. These two aircraft remained at Wright Field for almost the majority of their service life and were used for flight test work. The remaining 28 BT-8's were delivered to Randolph Field, the chief Army training field. All BT-8's were delivered by the end of 1936.

The Seversky BT-8 was well received by both student and Air Corps instructors where the majority were based. Randolph Field was known as the "West Point of the Air." It was here that many of our most notable Army aviators were trained. It was here also that the BT-8 proved so popular with their warm enclosed type cockpits. Pilot comfort was at last at hand. Previous type training aircraft were open cockpit biplanes. As the BT-8 continued on in its training

1st Air Corps contract was for 30 Seversky BT-8 Basic Trainers.

Seversky BT-8 on cross country flight to Barksdale Field.

role, other competing aircraft rose up to challenge the Seversky BT-8. One of the more successful competitors was the North American BT-9. It too was of advanced design and fulfilled the role as a modern basic trainer.

The BT-8 was a well built all metal aircraft, however it had a few pitfalls. The BT-8 was underpowered. Being of all metal construction, it was naturally heavy and the Army supplied R-985-11 only gave 400 h.p. Seversky pleaded with the military brass in Washington to let him install a more powerful engine but the Army had a hard and fast rule restricting training aircraft to a 400 h.p. limit. Pilots who knew how to handle the BT-8 had no problems, but if one buzzed the field and pulled up to abruptly he could stall out. If full power was lost under 500 feet the resulting crash generally proved fatal, for there was insufficient altitude in which to recover the stalled aircraft. The narrow landing gear gave some student cadets a problem, for if one was not sharp he could easily ground loop the aircraft. A good pilot held it straight and had no such problems.

By 1938 the lighter BT-9 was favored over the BT-8 in the training role. In early 1939 the majority of BT-8's were transferred from basic training role to primary training at Kelly Field. While at Kelly, they were also used for navigational work and radio and communication training. The last Randolph Field class to use BT-8's was in mid 1939. Shortly thereafter the Army ordered all BT-8's scrapped. The engines, instruments and other valuable parts were salvaged and the rest cut up. Not one is left today.

Advance design of BT-8 made all previous trainers obsolete.

SEVERSKY X-BT

Before the BT-8 Basic Trainer contract was completed, Maj. Seversky and Alex Kartveli designed the X-BT. Maj. Seversky knew the BT-8 was sadly underpowered and desperately tried to convince the Air Corps officials to abandon its stipulation that a 400 h.p. engine be used. His pleas fell on deaf ears.

The X-BT was built in 1936 to prove his theory on what the basic trainer could do and to show the great potential it had with these new improvements. The X-BT bore the civil registration number X-139M. Although Maj. Seversky described it as being in the basic training class, it was far more powerful and was more closely related to the Seversky 2XP.

The X-BT was a two place, low wing, all metal aircraft with a retractable landing gear. It mounted a Pratt-Whitney S3H-1 "Wasp" engine of 550 h.p. A two bladed Hamilton Standard propeller was fitted. It had wing dihedral and mounted a radio antennae mast. The tail wheel did not retract.

The X-BT was demonstrated to Army Air Corps at Wright Field competition in 1936, but it lost out to the North American BT-9. It was later demonstrated throughout the United States. During this time Maj. Seversky began to look overseas for orders for this unique airplane. He desired to sell this aircraft as a dual aircraft. By interchanging the removable outer wing panels from 36 to 41 feet one could have a long range escort fighter or a short range fighter or trainer. All the doors to him were being closed at the War Department for this airplane. Maj. Seversky was a business man. He knew that if they turned deaf ears toward him in Washington there were greener pastures overseas. Maj. Seversky's demonstration flights left a marked impression on all who witnessed these flights.

II SEVERSKY SPEED SEEKERS

FRANK FULLER

In 1936 Maj. Seversky planned to enter the Bendix Trophy Race. Air Corps officials said, "NO!" They feared his civilian Model 1XP contained too many military secrets. He had expected this. He felt this was not the real reason. They (War Department) held a grudge against him, for Maj. Seversky was a supporter of Gen. Billy Mitchell, and had been an advisor to General Mitchell when he was conducting his bombings of the Battleships in 1921 off the Virginia Capes. Maj. Seversky was a critic of Army brass and when he felt right was on his side, he had to speak out.

In 1937 Maj. Seversky contacted Frank Fuller, a sportsman pilot from California and President of the Fuller Paint Company. He inquired if he would be interested in piloting one of his Seversky racers in the forthcoming Bendix Race. Maj. Seversky reasoned that if Army brass in Washington would not let him fly his own aircraft, then American civilian pilots could receive approval and compete in this all important air race.

Frank Fuller accepts delivery of his racer from Maj.Seversky. Note straight wing of aircraft; dihedral was later added to wing.

Frank Fuller in cockpit of his SEV-S2 Bendix Racer. Fuller won 1937 & 1939 Bendix Trophy race in this aircraft. He placed 2nd in 1938 race.

Frank Fuller accepted Maj. Seversky's gracious offer and visited the Seversky factory on Long Island. He was impressed. There was not another aircraft in the world that could surpass it, and winning the Bendix Trophy Race would show what a really fast airplane the Seversky racer was. Air Corps permission was given for Frank Fuller to enter a Seversky in the race, provided he did not take the aircraft out of the country or permit foreign pilots to fly it. This year, Maj. Seversky also received permission for he and his company test pilot, Frank Sinclair to enter the 1937 Bendix Race. On September 2, 1937, a few days before the race Maj. Seversky flying Seversky AP-2 Registration No. R-1250 developed landing gear trouble and the aircraft was badly damaged in a belly landing at Floyd Bennet Field. Maj. Seversky's plane was out of the race. Fuller and Sinclair flew into Burbank, California without trouble. Sinclair flew the Seversky 1XP Registration No. R-18Y. Fuller flew his new Seversky SEV S-2 "Special Executive" Registration No. R-70Y. Both of these aircraft were in natural metal finish.

Ray Moore flew Fullers Seversky S2 racer in 1937 Thompson and placed sixth.

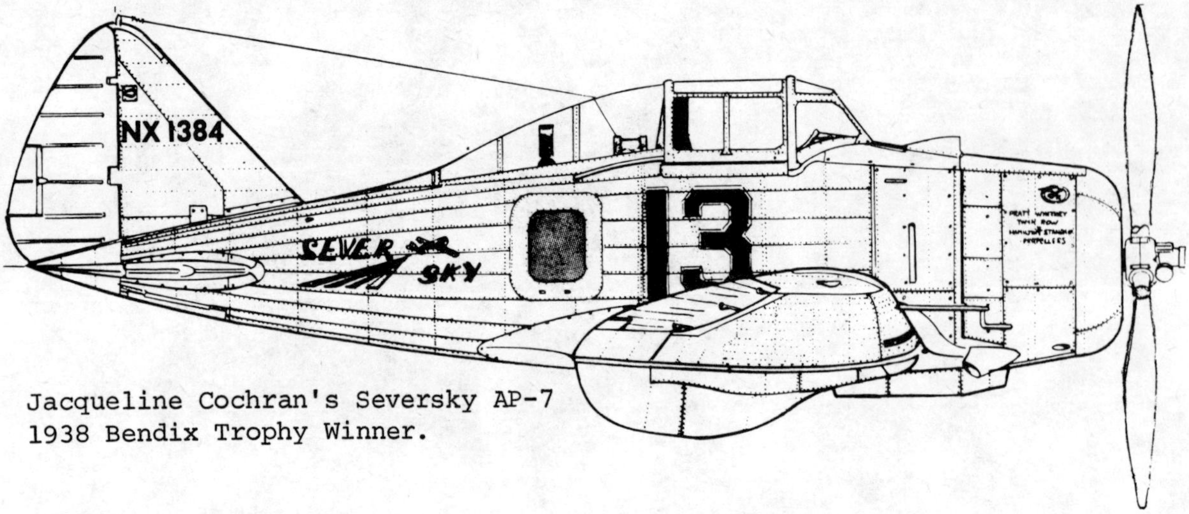

Jacqueline Cochran's Seversky AP-7
1938 Bendix Trophy Winner.

Seversky 1XP as flown by Frank Sinclair in 1937 Bendix Trophy Race. This aircraft placed 4th in 1937 Bendix and Thompson Trophy Races.

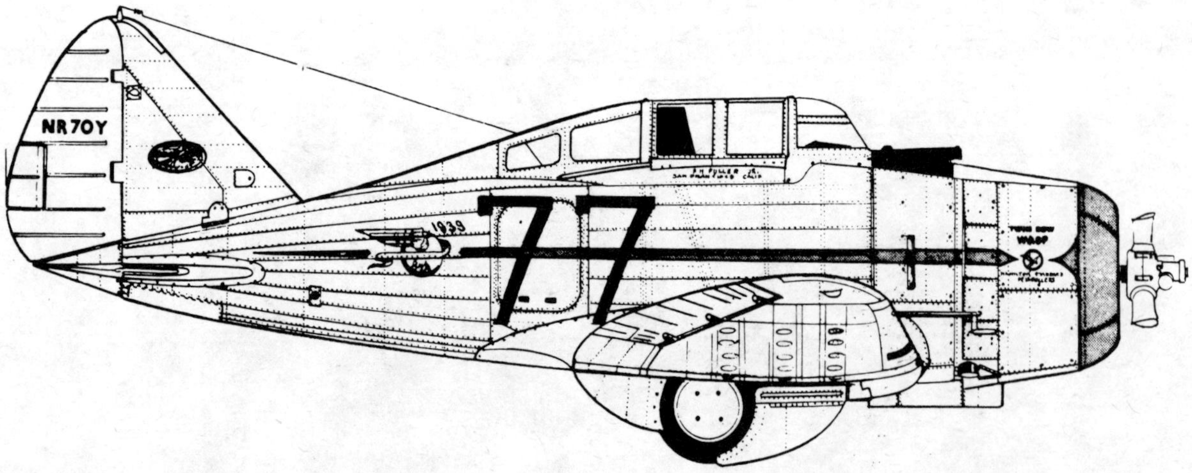

Frank Fuller's Seversky S 2 which placed 2nd in 1938 Bendix Trophy Race.

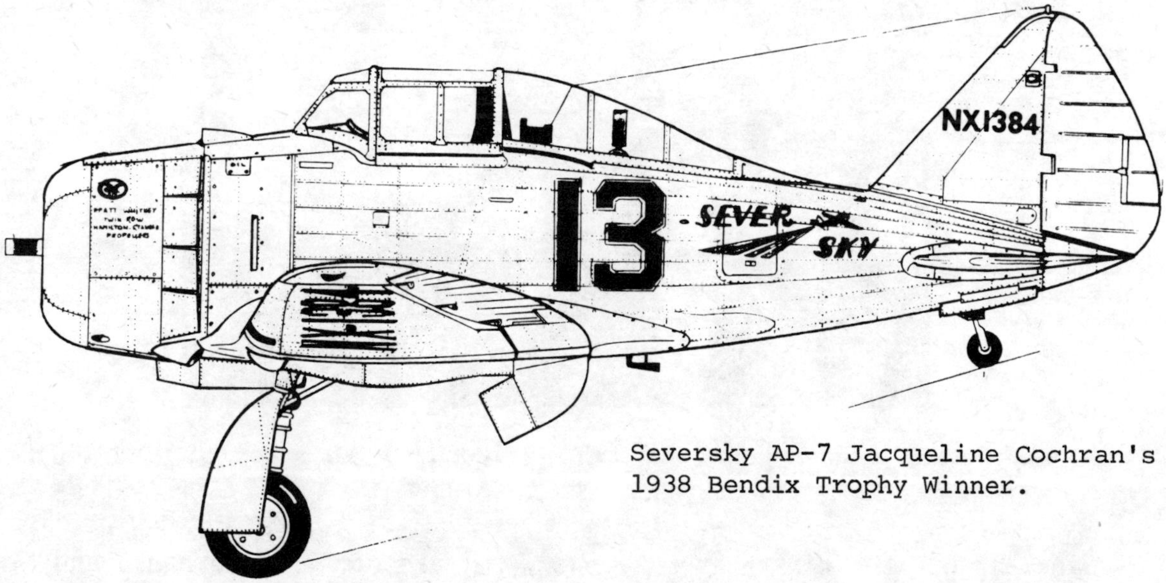

Seversky AP-7 Jacqueline Cochran's 1938 Bendix Trophy Winner.

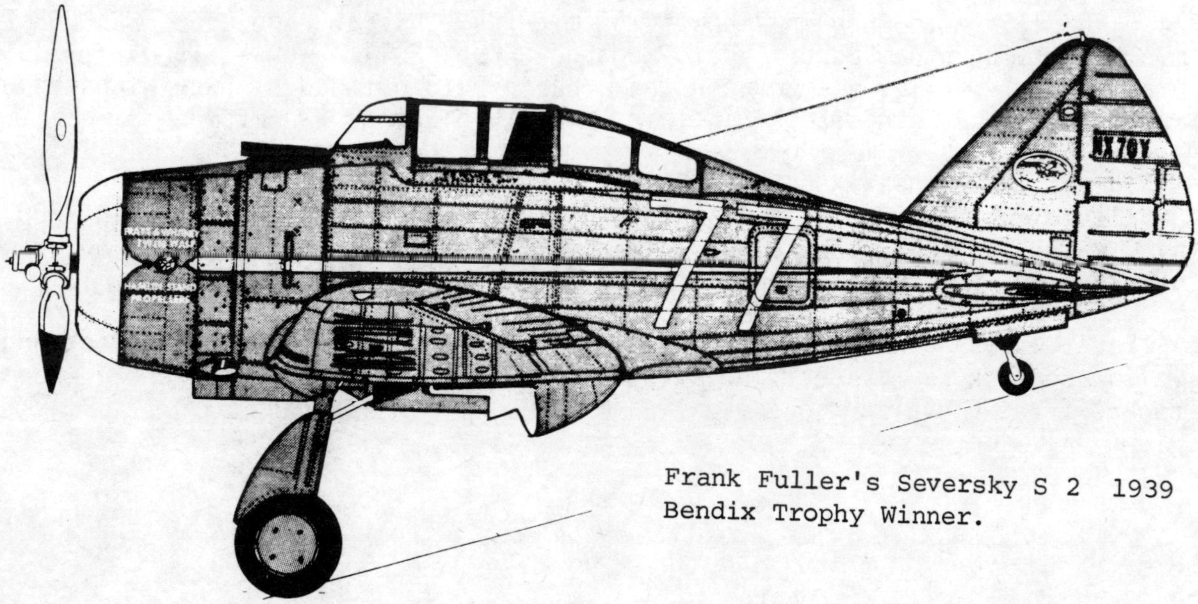

Frank Fuller's Seversky S 2 1939 Bendix Trophy Winner.

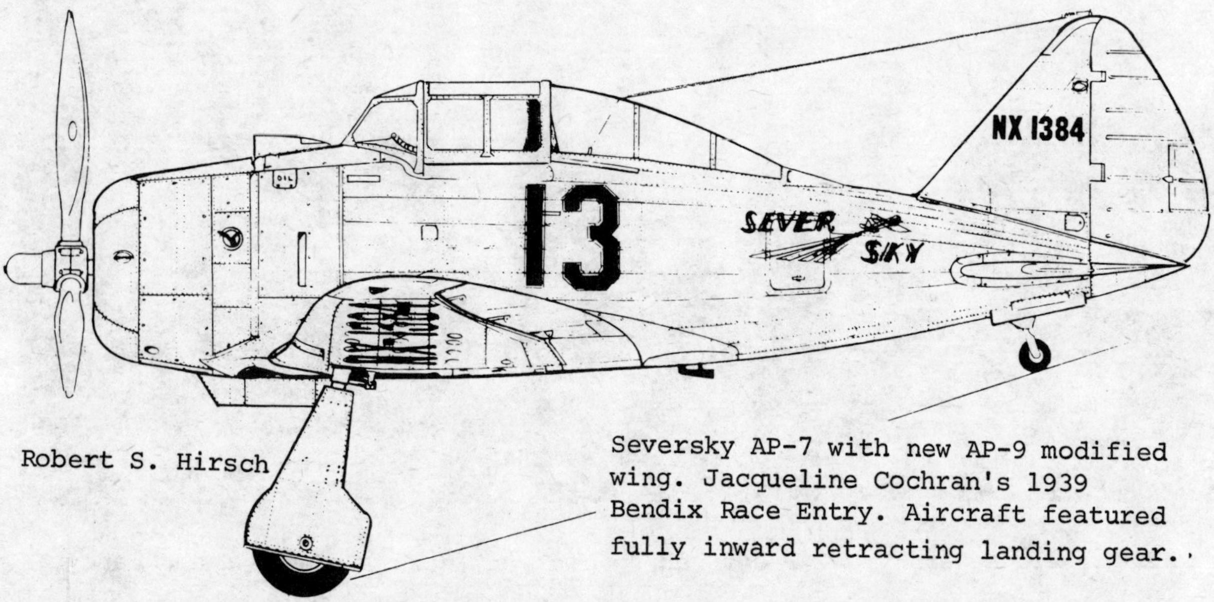

Robert S. Hirsch

Seversky AP-7 with new AP-9 modified wing. Jacqueline Cochran's 1939 Bendix Race Entry. Aircraft featured fully inward retracting landing gear.

Frank Fuller's SEV-S2 #23 won 1937 Bendix Race at speed of 258 m.p.h.

Sinclair was the first Seversky entry to take off from Burbank, and Frank Fuller was right behind him. Fuller flew a perfect race, crossing the finish line in Cleveland, Ohio in 7 hours, 55 minutes. His average speed was 258.2 m.p.h. setting a new Bendix Speed Record. Frank Fuller won $14,500, for 1st place money. Fuller's S-2 racer was a basic P-35 but had a lower windshield and canopy to lessen drag. Sinclair came 4th place at 184 m.p.h.

In the 1938 Bendix Race Frank Fuller, flying his Seversky S-2 racer developed engine trouble and was forced to reduce the power of his big 1200 h.p. R-1830 engine. This slowed him down but he did manage to take 2nd place at 238.6 m.p.h. and collect a purse of $5,800 dollars.

For the 1939 Bendix Race Frank Fuller entered his newly painted green and yellow Seversky S-2 racer, and at 3:00 AM on a very foggy September 2, 1939 took off for Cleveland, Ohio.

Frank Fullers Seversky S-2 racer had special cut down windshield and canopy.

Frank Fuller's newly painted green & yellow Sev-2S racer impounded at Burbank, Calif.,for start of 1939 Rendix Air Race. Note new race #77.

Frank Fuller crossed the finish line at Cleveland in 7 hours, 14 minutes, 19 seconds, at an average speed of 282.1 m.p.h. setting a new Bendix record. He collected $12,500 for 1st place. On the day before the air races at Cleveland started, German military forces invaded Poland, September 1, 1939. Maj. Seversky realized that the 1939 Cleveland Air Races would be the last until the war in Europe was over.

In early October 1939 Seversky was off to Norway to demonstrate his Seversky EP-1 to the Norwegian Government as a possible replacement for its aging fleet of fighter aircraft. On October 12, 1939 while he was in Norway Maj. Seversky was voted out of office by the company board of directors. The firm was renamed Republic Aviation Corporation.

A joyus occasion: Gar Wood (Rt) presents record breaking trophy to Miss Jacqueline Cochran while Maj. Seversky looks on. Aircraft is SEV-1XP.

Jacqueline Cochran won the Gar Wood Trophy in the Seversky 1XP.

JACQUELINE COCHRAN "AIRBORN CINDERELLA"

Jacqueline Cochran was one of America's greatest women pilots. She was born in Pensacola, Florida and learned to fly at Roosevelt Field, New York in December 1932. She flew her first air race in 1934, London to Melbourne, but had to drop out over Rumania because of engine problems. In 1935 she became the first woman to enter the Bendix Trophy Race. It was not until 1937 when she was given the opportunity by Maj. Seversky to fly one of his Seversky racer aircraft that she acquired world wide notice.

War Department officials complained to Maj. Seversky that it was "undignified" for the President of a major aviation company to pilot planes and set speed records. In order to overcome these absurd objections Maj. Seversky invited a relatively unknown girl flyer a chance to fly one of his sleek, speedy aircraft.

On April 6,1940 J.Cockran set a new 2000 km. speed record of 332 mph.

Jacqueline Cochran gladly accepted Maj. Seversky's offer. She received a complete checkout and after several trial flights, she started to blaze a trail of intercity speed records. In September 1937 flying the veteran Seversky 1XP, civil registration R-18Y she set a new women's speed record of 293 m.p.h. over a 3 km. course at Detroit, Michigan. This was the same airplane that Frank Sinclair flew in the 1937 Bendix Race. On December 3, 1937 she established a new non-stop New York to Miami speed record of 4 hours, 12 minutes at an average speed of 300 m.p.h. Ten days later she set a new national women's speed record of 255.9 m.p.h. over a 100 km. course.

Maj.Seversky & Miss Cochran by SEV-1XP. She set womens speed record of 293 m.p.h. at Detroit, Michigan in this aircraft September 1937.

Jacqueline Cochran won the 1938 Bendix Trophy Race in this Seversky AP-7.

For the 1938 Bendix Race Maj. Seversky had a new Seversky racer for her to fly. Known as the AP-7 Registration No. NX-1384, this new aircraft was powered by a new Pratt-Whitney R-1830 of 1200 h.p. Jackie flew into Burbank where her Seversky AP-7 and all other entries were impounded before the race. On race day she was the third aircraft off the ground. Her race plan was to fly high at 20,000 feet and use oxygen. A short while out she developed a plugged fuel line, but by rocking the wings she found she could keep the engine running. It was later found that debris had somehow gotten into her fuel tank and covered the fuel outlet line to the engine. Her rocking motion of the aircraft uncovered the outlet and permitted fuel into the engine without interruption. She crossed the finish line in Cleveland in 8 hours, 10 minutes, 31 seconds at an average speed of 249.7 m.p.h. Jacqueline Cochran continued on to Bendix, New Jersey to set a new West-East speed record. Maj. Seversky was elated. Two of his aircraft had won the Bendix Race two years in a row, and they had finished one, two in the 1938 Bendix which was not a bad record.

Miss Cochran's Seversky AP-7 1939 Bendix Racer with newly added inward retracting landing gear. Bad weather prevented her from competing in race.

Maj. Seversky congratulates Jackie Cochran on hew winning Gar Wood Trophy.

During this period flying Seversky aircraft she became known as one of the fastest women pilots in the world. She was always a fine competitor and will go down in history as one of America's two greatest women pilots. Maj. Seversky was delighted with her racing successes and he provided her with a newly rebuilt AP-7 for the 1939 Bendix Race. This aircraft had a new inward retracting landing gear, a new flush riveted wing with a sharper leading edge.

She developed landing gear problems heading west for the Bendix Race and had to land at Dayton, Ohio where repairs were made. After being laid up for several days she headed West once again to compete in this greatest of all transcontinental air races. Fortune did not smile upon her this year, for she made two attempts to take off but each time she neared the starting line dense fog would close the airport forcing her return to the hangar. She asked for permission of air race officials for a time extension, but this was refused and rather than risk losing an aircraft she did not own and possibly her life she wisely abandoned the race. On April 6, 1940 she set a new women's 2000 km. speed record of 332 m.p.h. piloting the Seversky AP-9.

Jacqueline Cochran in SEV-1XP in which she set many intercity speed records. Winning combination of Miss Cochran & SEV-1XP racer brought national attention to women pilots in the United States.

During World War II she became founder and director of the Women's Air Force Service Pilots organization which was better known as "WASPS." She became the first woman pilot to exceed the speed of sound. Her record of accomplishments are long and many. She will go down in the annuals of American Aviation as one of the greatest American Woman pilots.

Piloted by Lee Miles in 1935 Thompson Trophy Race the SEV-3M placed fifth.

SEVERSKY RACERS

The first Seversky airplane to enter a Thompson Trophy Race was in 1935 when Lee Miles piloted the all gold Seversky SEV-3M Amphibian, Registration No. X-2106. This marked the first time that an amphibian had been entered in the Thompson Race. Lee Miles placed 5th in a field of six airplanes entered. He averaged 193.5 m.p.h. for this closed course race.

Jacqueline Cochran & Maj. Seversky pose by AP-7 in which Miss Cochran won the 1938 Bendix Trophy Race. She won Bendix purse of $ 12,500.00

Seversky 1XP just prior to 1937 Thompson Trophy Race. During its racing career at the National Air Races the SEV-1XP bore the race #63. Note full pants which covered the landing gear & decreased drag. When not racing, aircraft served as Maj.Seversky's personal aircraft, & carried passenger in rear. Note passenger window in right side door.

Seversky 1XP racer originally had straight wing. Note their is no dehedral.

1st Seversky aircraft Jacqueline Cochran flew was SEV-1XP. She was a trailblazer and became the fastest women aviatrix in the world.

After Army approval had been given for Seversky participation in the 1937 Cleveland Air Races Maj. Seversky entered his Model 1XP, Registration No. R-18Y in both the Bendix and Thompson Race. Frank Sinclair was to be the pilot. Also entered in the 1937 Thompson Race was Frank Fuller's Seversky S-2 to be flown by Ray Moore. Both of these racers were powered by the powerful Pratt-Whitney R-1830 engine of 1000 h.p.

Frank Sinclair took 4th place in the Thompson at an average speed of 252.3 m.p.h., Ray Moore came in 6th at a speed of 238.4 m.p.h. After the 1937 air race, officials at Cleveland would not permit a Bendix Race entry to compete in the Thompson Trophy Race, therefore there were no Seversky aircraft entered in the 1938 or 1939 Thompson.

The Thompson Trophy

Seversky 1XP was flown by Frank Sinclair in 1937 Bendix & Thompson Races. He placed forth in each race. The SEV-1XP was a record breaking aircraft.

This war surplus AT-12 was flown by Vince Perron in 1949 Bendix Trophy Race.

The last Seversky design to enter a Bendix Race was a Seversky 2PA-204A. This aircraft was originally purchased by Sweden, taken over by the U. S. Government and designated AT-12 "Guardsman." This aircraft was owned by movie actor Buddy Rodgers and named "Buck Rodgers." It was flown by Vince Perron, a local Southern California flight instructor. It was entered in the 1949 Bendix Race. This race was the first Bendix Race to have a "race horse" start, where all the race entries took off all at once. A wide area was needed for this takeoff and so the takeoff field was moved to Rosamond Dry Lake hear Lancaster, California, on the Mojave Desert. There were just six entries this year. Pilot Perrone knew that his racer did not have a chance to win but all he had to do was to finish the race and he would collect $1,000.00 for 6th place money. Perrone was late getting off the dry lake bed, for he had engine trouble. He made it to Grand Junction, Colorado where he was forced to land and was out of the race, however he did win $750.00 consolation prize money.

Movie stars (l-r) Buddy Rogers & Mary Pickford wish AT-12 race pilot Vince Perron good luck before 1949 Bendix Race.

The AT-12 was one of six entrys in 1949 Bendix Race. AT-12 dropped out of race over Colorado due to engine problems.

Jimmy Doolittle flew this SEV-DS "Doolittle Special" for Shell Oil Co.

The Doolittle SEV-D S proved to be a dynamic advertisement for Maj.Seversky's company. Note prop spinner was later added to improve speed & better cooling.

JIMMY DOOLITTLE AND THE SHELL-SEVERSKY D S

"DOOLITTLE SPECIAL"

The Shell Oil Company purchased a special Seversky D S airplane for Jimmy Doolittle. This aircraft bore Registration No. NX-1291. This fast executive aircraft was ordered by Shell in 1936 and was delivered in 1937 and based at St. Louis, Mo.

Jimmy Doolittle had retired from active air racing, however he continued flying and testing aircraft for many, many years. He had been flying the company's Lockheed "Orion," a fast, low wing monoplane transport for several years. With the coming of the modern, all metal Seversky Pursuit, Doolittle made arrangements for Shell Oil Company to buy one for his personal use and also to experiment in the tests of 100 octane fuel which Shell Oil was developing.

The development of 100 octane fuel paid great dividends to our Air Force during World War II, for it was a better grade of aviation fuel than either the Japanese or German variety.

Jimmy Doolittle's Seversky D S featured a straight wing and mounted a large Wright R-1820 engine of 850 h.p. It was kept in its natural metal finish.

As an Army Reserve officer Doolittle returned to Air Corps service in 1940. It is believed that his aircraft was later sold overseas by Shell Oil Company.

Shell Oil Co. purchased Seversky D S in 1937. It was one of two civilian Seversky models sold in the United States. It powered by Wright R-1820 engine. Besides testing aviation fuels, Doolittle Seversky D S was used as a fast transport aircraft. Flying Shell executives coast to coast for important business meetings it can claim to be the 1st "corporate aircraft" in the United States. Note passenger window just below pilots canopy.

Jimmy Doolittle flew this Seversky D S (Doolittle Special) for Shell Oil Co. Doolittle used this aircraft testing Shell 100 octane fuel.

This "Drake Special" racer was actually Frank Fullers SEV-S 2 made up for M.G.M. movie "Test Pilot" which starred Clark Gable.

The Seversky pursuit design was used by Hollywood Studios for many years as enemy fighter planes, mystery fighters and racing planes.

In 1938 M.G.M. Studios in Culver City, California rented Frank Fuller's Seversky S 2 Bendix Trophy Winner for the motion picture "Test Pilot" starring Clark Gable, Myrna Loy, and Spencer Tracy.

In this picture the Seversky S 2 racer was erroneously labeled the "Drake Bullet," and this name was applied to both sides of the aircraft. It bore the fictitious civil Registration No. R-23655. M.G.M. Studios also made a complete non-flyable mock-up of the Seversky aircraft. This Seversky mock-up was used in several other M.G.M. motion pictures: "Too Hot To Handle," again starring Clark Gable and Myrna Loy, and "Flight Command," starring Robert Taylor.

Wood models of the Seversky P-35 design were built by Warner Brothers Studios. They were used in one of their greatest World War II productions titled: "Air Force," starring John Garfield, and Harey Carey. In this picture they were cast as Japanese fighter planes and one of the more stirring moments of this film comes when John Garfield shoots one of these aircraft down, holding a .50 caliber machine gun by hand and firing it from the hip.

In 1958 M.G.M. made a motion picture titled: "Never So Few," starring Frank Sinatra, Gina Lolabrigeta, and Steve McQueen. In this picture the original studio mock-up was rolled out, dusted off and painted like a Japanese fighter plane. It was used in an airfield scene along with an original A6M-5 Zero Fighter and a Japanese KI-84 "Hayate" Army fighter which was taxied down the runway.

Three of the most outstanding motion picture stars M.G.M Studio's had under contract were Clark Gable, Myrna Loy, and Spencer Tracy shown here by tail of Seversky S2 racer for M.G.M. motion picture "Test Pilot."

Originally built for MGM's film "Test Pilot" this studio mock-up of Seversky aircraft was repainted as Japanese fighter and used in MGM's film "Never So Few"

III P-35 "THE FARMINGDALE FLASH"

The first production P-35 was sent to Wright Field where it served in flight test work. It was there from 1936 to 1941. The remaining production P-35's were sent to 1st Pursuit Group at Selfridge Field, outside Detroit, Michigan. Approximately 25 P-35's were assigned to each of the three squadrons which formed the 1st Pursuit Group. These three squadrons were 17th "Snow Owl," 27th "Diving Hawk," and 94th "Indian Head," Pursuit Squadrons. They operated P-35's from 1938 to 1940. They also flew during this same period the Curtiss P-36 "Hawk." The Seversky P-35's were flown in the morning and the P-36's in the afternoon. These two aircraft divided the Air Corps into two camps. The argument of which aircraft was a better pursuit caused hatred to build and consequently what the Air Corps thought would be a good idea proved to be a tragic error.

P-35 S/N 36-355 was used as a test aircraft at Wright Field from July 1937 to January 1938. The Material Division tested P-35 S/N 36-408 from August 1939 to February 1940 when it was lost in an accident.

P-35 Nameplate

Top speed of P-35 was 281 m.p.h.

One of three P-35 test aircraft at Wright Field and flown by the noted Air Corps test pilot Capt.Stanley M.Umstead. Note Wright Field Insignia on body.

1st production P-35 delivered to Air Corps had fully enclosed landing gear. This feature was later deleted on all futher production P-35 models.

Full head on view of 1st production P-35 which shows it delivered to Wright Field with straight wing. Dihedral was later added to all P-35 models built.

P-35 was powered by P&W R-1830-9 of 950 h.p. It had a top speed of 281 m.p.h.

This rear view shows off-set rudder to counter engine torque.
Note dehedral has been added to wing.

Seversky 2PA "Convoy Fighter" was designed to be a jack of all trades aircraft. It could be used as a Basic Trainer, Advanced Trainer, a Convoy Fighter, an Observation Aircraft, an Attack Bomber, and a Light Bomber. It was one of the worlds fastest 2-seat aircraft. This one shown above went to Russia.

Seversky 2PA-B3. Twenty of these aircraft were exported to Japan for use by the Imperial Japanese Navy where they were designated A8V1 and were used in operations in China prior to Pearl harbor. Several of these aircraft were later given to the Asahi Shimbun Newspaper where they were used as fast courier aircraft.

Seversky P=35 of 17th Pursuit Sq., 1st Pursuit Group at Selfridge Field, Michigan. Note black section leader stripe on fuselage. Tail marking PA 16 denotes 16th Aircraft of 1st Pursuit. The personnel of the 17th Pursuit Sq. were deployed to the Philippine Islands in late 1940. They left their P-35's at Selfridge Field but received later P-35A's which had been seized from shipment to Sweden.

P-35 was lightly armed, having only one 50 and one 30 caliber machine gun.

Line up of 27th Pursuit Sq. P-35's at Selfridge Field, Michigan.

Seversky completed and delivered the last ten P-35's in August 1937. While at Selfridge Field, P-35's were flown in natural metal finish. They were the first Army Air Corps pursuit aircraft to enter service without the traditional yellow and blue Air Corps paint scheme. P-35's carried standard Air Corps markings including group designators and aircraft numbers on cowl and leader color bands. Headquarters used numbers 1 to 10; numbers 11 to 39 were assigned 17th Pursuit Squadron aircraft; numbers 40 to 69 were assigned 27th Pursuit Squadron; and 70 to 99 were assigned to 94th Pursuit Squadron.

17th Pursuit Sq. pilot prepares to start up P-35.

1st Pursuit Group Headquarters P-35 at Selfridge Field.

During service with the 1st Pursuit Group 1938 to 1940 nineteen P-35's were wrecked or lost through accidents. The P-35's averaged 900 flying hours during their Air Corps service careers at Selfridge Field. One P-35 S/N 36-707 hit an all time high of 1490 hours before it was scrapped in November 1942.

In December 1940, 17th Pursuit Squadron personnel were ordered to Nichols Field, Philippine Islands, for a two year tour of overseas duty. The Air Corps was beginning phase one of defense of the Philippine Islands and Air Corps expansion was in high gear.

Impressive line up of American P-35's of 1st Pursuit Group.

P-35 aircraft of 27th Pursuit Sq. were allocated numbers PA 40 through PA 69.

While at Selfridge Field the three pursuit squadrons comprising the 1st Pursuit Group were color coded: aircraft of the 17th squadron had white color cowls, the 27th squadron had yellow cowls, and the 94th had red cowl bands.

1st Pursuit Grp. headquarters P-35. Note group insignia on side.Nose cowl was painted in three segments: Red, Yellow,& White colors denoting each sq.

"Aerial Artistry" P-35's of 27th Pursuit Sq. in line abreast formation.

Aircraft of 94th Pursuit Sq. were allocated numbers PA 70 through PA 99.

Two yellow fuselage bands around this P-35 denotes squadron leader of 27th Pursuit Sq.

Line up of 1st Pursuit Group P-35's at Selfridge Field.

During 1940 war maneuvers at Ft. Bragg, North Carolina, P-35 aircraft of 27th and 94th Pursuit Squadrons were camouflaged in water base paint of sand, olive drab, and green. Some were gaudy multi-colors of different hues and used as an experiment. This experimental camouflage paint was used only during the war games and the aircraft were later restored to natural metal finish.

Water color camouflage painted on 94th & 27th Pursuit Sq's. P-35 for 1940 war games. Note barely discernable "Indian Head" insignia on fuselage.

1st Pursuit Group cadred the 31st Pursuit Group at Selfridge Field of which this P-35 was attached.

During the expansion program of 1940 to 1941, five additional pursuit groups were formed at Selfridge Field: the 31st, 49th, 50th, 52nd, and 58th. All trained on Seversky P-35 aircraft during these peaceful pre-World War II days. By December 7, 1941 fifty-one P-35's were still in active service, however their active career was rapidly coming to an end. More modern equipment was replacing these aging fighters as rapidly as possible. By mid-1942 most of the remaining P-35's were being assigned to training schools for use as instructional machines for mechanic trainees. By this time some 30 P-35's had accumulated over 1000 flying hours. By June 1942 there were only four P-35's remaining and by "VJ-Day" August 8, 1945 only one P-35 remained. This last P-35 S/N 36-404 survived in a training school in Minnesota.

Seversky P-35 had Clark YH airfoil for rapid climb, good speed and lift requirements.

SEVERSKY NF-1 NAVY PURSUIT

In early 1937 Major Seversky proposed to the U.S. Navy Department that they test fly one of his company's Seversky pursuits for possible carrier use without cost to the Navy. The U.S. Navy accepted Seversky's gracious offer and on September 24, 1937 the Seversky NV-1 was delivered to Annacostia Naval Air Station. The NF-1 bore the Registration No. X-1257; it was a company owned aircraft. The Navy Department realized that the low wing fighter was the design of the future, for they cancelled the biplane Grumman XF4F-1 design in July 1936. Testing the NF-1 would make a good comparison with current Navy pursuit models.

The NF-1 had a wing span of 36 feet and a length of 25 feet 2 inches. It was a basic P-35 airframe with some refinements. It mounted the Wright R-1820-22 air cooled, radial engine of 950 h.p. It was equipped with an arresting hook for carrier tests and mounted a small bomb rack on the bottom of the wing. The NF-1 had a straight wing, and a retractable landing gear and tailwheel. This aircraft was easily identified by its bulged windshield. It had a maximum speed of 267 m.p.h. at 15,200 feet and had a service ceiling of 30,700 feet. Navy trials showed it to have a landing speed of 69 m.p.h., but the performance was not considered an advancement over the current fighters then in Navy service. The NF-1 flight test data was used by the Navy to compare the later Grumman F4F and Brewster F2A carrier fighters tested by the Navy. The Brewster entry won the initial contract in the design competition. The NF-1 was returned to the factory and this concluded further participation by Seversky into Navy competition.

U.S. Navy tested NF-1 model in 1937 but landing speed was considered to high for carrier use.

Navy requested Maj. Seversky to change windshield on NF-1 as pilots complained of poor visibility. This view shows the aircraft with standard P-35 windscreen with Navy gunsight.

U.S.NAVY

Seversky NF-1 as it first appeared for Navy test at Anacostia N.A.S. 1937. Note early style bulge windscreen and carrier arresting hook.

SEVERSKY SUPER CLIPPER

In late 1937 Maj. Seversky had Alex Kartveli design a new revolutionary 120 passenger, 300 m.p.h. flying boat. Maj. Seversky was anxious to secure a contract for this commercial flying boat to diversify his aircraft company from building strictly military airplanes. Douglas, Boeing, and Curtiss had all built successful commercial transport aircraft and Maj. Seversky was very determined that his company would also enter this lucrative market.

Maj. Seversky hoped to win a contract from Pan American Airways for from 3 to 12 of these transoceanic flying boats. This design was known as the Seversky Super Clipper. It would carry a 43,000 pound load, cruise at 250 m.p.h., and have a top speed of 300 m.p.h. It was an all metal flying boat, had a wing span of 250 feet, mounted eight liquid cooled engines of 2000 h.p. each and was to carry 120 passengers. This was a very revolutionary flying boat design and was unique in many ways: The passengers would be housed entirely within the wing from tip to tip, in generous staterooms. The living quarters would be pressurized for passenger comfort at altitudes above 15,000 feet. This new flying boat design employed two large fuselages, or outriggers, which housed retractable landing pontoons and supported the tail surfaces. The retractable landing pontoons were a patented Seversky feature. These pontoons absorbed the shock at takeoff and landing and being fully retractable increased the flying boat's air speed after it took to the air.

The eight liquid cooled engine installation was unique: The two power plants mounted forward and one aft were each comprised of two engines coupled to a single propeller. This decreases torque and permitted use of a large diameter propeller. Mounted at the wing trailing edge were two more powerplants, each equipped with a single engine.

Each liquid cooled engine was rated at 2000 h.p. and able to develop 2300 h.p. for takeoff. Each of these engine nacelles were large enough to allow two mechanics to work on the engine at one time while in flight.

The Seversky Super Clipper had a fuel capacity of 17,000 gallons. De-icing equipment was to have been installed. Flaps of the latest Seversky-Kartveli design were included to insure slow landing speed and accelerated takeoffs. This flying boat was to be constructed of spot welded stainless steel. This new construction process was developed by the Edward G. Budd Manufacturing Co.

Maj. Seversky claimed that this flying boat could be converted to a heavy bomber and could carry ten 2000 pound bombs, 12,000 miles to Europe and back at a speed of 300 m.p.h. Maj. Seversky's bomber concept of using the United States as the main bombing bases from which to make long range attacks was too advanced in thinking for military brass in 1937-1938. Eight to nine years had to pass before Air Force brass began using the B-36 global bomber utilizing Maj. Seversky's bomber concept.

Pan American Airways considered Maj. Seversky's proposal but ordered more conventional Boeing Model 314 "Clipper" seaplanes for their worldwide airline. No other market prospects appeared and the design was dropped and no aircraft were built.

Scale model of the giant Seversky Super Clipper for Pan American Airways. This aircraft was never built.

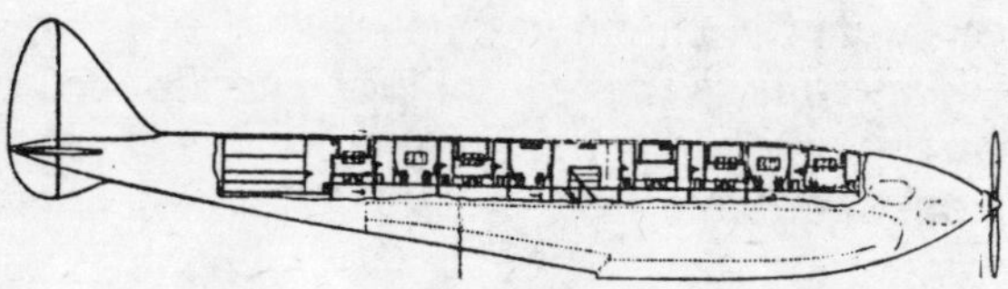

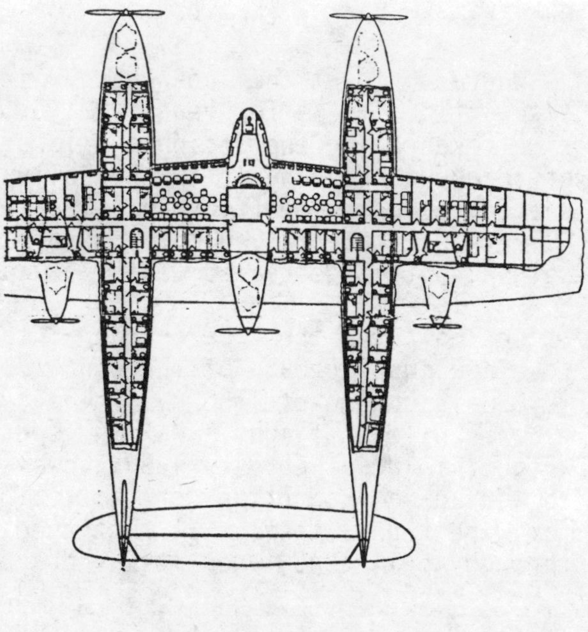

XP-41 Wright Field test aircraft at Langley Field, Virginia. Only one was built.

After delivery of the last production P-35 aircraft to the 1st Pursuit Group, Major Seversky and his chief designer Alex Kartveli proposed to the Air Corps that they modify the last production P-35 S/N 36-430 with a new wing center section, a fully retractable landing gear and a turbo supercharger in the belly of the fuselage aft of the wing. Approval was granted by the Air Corps. This XP-41 design paralleled the design and construction of the company's project, the Seversky AP-4. Both aircraft employed General Electric turbo superchargers. The AP-4 Reg. No. NX-2597 also employed a fully retractable inward landing gear and from all outward appearances looked almost identical to XP-41, however the AP-4 used a 3 blade Hamilton Standard propeller. The supercharger location was the same as in the XP-41.

Where to mount the supercharger in a radial air cooled fighter had plagued aircraft designers for years. Major Seversky and Alex Kartveli spent considerable time researching the design and location of the supercharger to ensure maximum efficiency. Placing it in the belly and away from the nose of the aircraft seemed to be the best possible location. Previous aircraft like the Consolidated PB-2A had located it on the side of the engine. This not only added extra nose weight, but it decreased visibility and increased drag.

On the XP-41 the supercharger intake was mounted on the left side of the fuselage just forward of the wing root. It had a 14 cylinder Pratt-Whitney R-1830-19 engine of 1150 h.p. The XP-41 was the first aircraft to exceed 300 m.p.h. in level flight. Maximum speed was 323 m.p.h. It was flight tested at Wright Field in February 1939, however it did not show sufficient improvement to warrant a production contract. The XP-41 was later sent to Langley Field for wind tunnel tests. It continued its remaining service life in test work through April 1942, when it was broken up and sold for scrap.

Seversky continued to experiment with his AP-4. It was a high altitude fighter. Seversky was a man of vision, a prophet in a land crying out to be prepared. His cries were ignored, but as a man of vision he realized that the next air war would be fought at high altitudes and that unless the Army Air Corps had such a high altitude fighter we would have a third rate air force. A turbo supercharger is necessary in a fighter to provide sufficient air to the engine at extreme altitudes where the air is very thin. Proper air-fuel mixture is necessary at all times to develop full power in military aircraft.

The AP-4 was similar to the Seversky AP-2 which Seversky was forced to crash land at Floyde Bennet Field, New York after he suffered gear failure in 1937. The AP-4 had inward retracting landing gear, flush riveted skin, and self sealing fuel tanks. It was tested at Wright Field in February 1939, and continued to undergo modifications to improve its design. When it arrived at Wright Field for the competition it looked like the XP-41 but by April 1939 a large pointed spinner and a very tight fitting cowl had been added. At first glance one would think it had a large liquid cooled engine but this was not the case. Alex Kartveli was experimenting in an attempt to clean up the drag caused by the large blunt radial engine mounted up front. In his attempt to improve streamlining, engine heat temps soared and further attempts to fully enclose a radial engine were dropped.

Major Seversky always favored radial air cooled engines over liquid cooled type engines. Less drag occurred with the latter engine, however they were not as reliable as the radial engine. Pressure was brought to bear on Maj. Seversky by the Air Corps for him to develop a liquid cooled fighter. A designation AP-3 was assigned to the development fighter but it was never built as Maj. Seversky devoted his full time to aircraft projects that held the greatest promise and he

Seversky AP-4 of 1939 featured inward retracting landing gear, turbo, and flush rivets.

Last P-35 built was modified to XP-41. Note inward retracting landing gear.

firmly believed in the air cooled engine fighter. His vision was later proved during World War II when P-47 Thunderbolt fighters returning from missions high over Germany had two to three engine cylinders completely shot off while the reliable P&W R-2800 soundly pounded its pilot safely home. All one has to do is read 56th Fighter Group Ace Maj. Robert L. Johnson's book "Thunderbolt" to realize this point.

Another Seversky entry in the February 1939 design competition was the Seversky AP-9 Reg. No. NX-2598. The AP-9 was reported to have been built from the Seversky AP-8. The AP-9 also employed an inward retracting landing gear and a more sharply pointed edge and was the second Seversky with flush rivet construction. It was powered by a Pratt-Whitney R-1830 engine of 1000 h.p. and had a reliable Hamilton Standard 3 bladed propeller. No turbo supercharger was mounted on the AP-9. None of the three aircraft Seversky entered received a contract order, for the Bell XP-39, and new XP-38 design held more promise. The AP-9 was returned to the factory where a more powerful 1200 h.p. R-1830 engine was installed.

On April 6, 1940 noted aviatrix Jacqueline Cochran set a Women's 2000 km. speed record in the AP-9 when she hit 322 m.p.h. The AP-9 was sold to the Dominican Republic in the summer of 1940.

Maj. Seversky set new transcontinental speed record in AP-7 of 10 Hrs.3 min. in September 1938. Two days later Miss Cochran won the 1938 Bendix Race in same aircraft. She hit an average speed of 249.7 m.p.h. during race.

Seversky AP-9 was entered in Wright Field design competition April 1939.

Jacqueline Cochran set speed record of 332 m.p.h. on April 6,1940 piloting this Seversky AP-9 pursuit. Only one AP-9 was built and flown.

Seversky AP-1 fitted with special cowl and pointed spinner to improve streamlining. Sharply pointed windscreen was later used on P-43 "Lancer" and early model P-47 "Razor Back." "Thunderbold" fighters

Profile view of Seversky AP-1 shows an extremely clean airframe design. Note A D F Loop antennae mounted directly under belly of plane. Tight fitting cowl of AP-1 created cooling problems and was discarded.

Seversky AP-1 is shown here with new cowl design. Frontal area has been enlarged to permit more cooling air to enclosed P&W R-1830 radial engine.

In an all out effort to clean up the basic P-35 design, Maj. Seversky in early 1938 had his chief design engineer, Alexander Kartveli cowl in the Pratt and Whitney R-1830 radial engine on the Seversky AP-1. The AP-1 was modified to give give a radial engine pursuit the streamlining of an inline engine pursuit. Less drag meant increased speed. Maj.Seversky was a firm believer in radial engines versus inline engine pursuits.

The AP-1 was a company owned aircraft. It was mainly used for flight test purposes. When it originally left the Long Island factory it had a bulged windshield, and a straight wing. It was flown to Wright Field prior to delivery of the first production P-35. It was later returned to the factory where a new wing with dehidral was added along with a standard P-35 windscreen. It was then flown to Langley Field, Virginia where wind tunnel tests were conducted. It later was returned to the factory.

Kartveli's design fully enclosed the R-1830 radial engine in a special tight fitting cowl. A large spinner was fitted which gave it a pointed nose. An oil cooler was also mounted. Since the aircraft was intended only to test cowling arrangements, no armament was fitted.

IV EXPORT MODELS 1937-1940

"SPEED WITH STAMINA"

Maj. Seversky shown flying the Seversky X-BT with Alex Kartveli in the rear seat. The X-BT was the forerunner of the Seversky "Convoy Fighter"

In 1938, Maj. Seversky had the XBT modified into a new design. A larger Wright R-1820 engine of 1000 h.p. was installed. Maj. Seversky had pioneered the use of interchangeable wing panels from 36 to 41 feet several years earlier, thus the wing area could be increased from 220 sq. feet to 250 sq. feet. He now proposed the introduction of this aircraft into a variety of military roles. The basic model was redesigned to accommodate different engines ranging from 400 h.p. to 1200 h.p. This allowed the aircraft to be used for multiple purposes from basic training to long range fighting to light bombing.

This feature of engine interchangeability was accomplished by designing the different engine installations as individual units which later became known in the Air Force as "Q.E.C.'s" or Quick Engine Change units, all of which were arranged to fit the same four main engine attach points at the fuselage firewall, including interchangeable fuel and oil lines, electric wiring and controls. By this unique technique a Wright R-1820 or P&W R-1830 engine could be installed at the buyer's preference and this two place fighter had an approximate speed of 300 m.p.h. Likewise the P&W R-1340 of 550 h.p. could be installed and the plane thus equipped became an observation scout plane. For basic or advance training purposes a P&W R-985 or Wright R-975 engine of 400 h.p. could be installed.

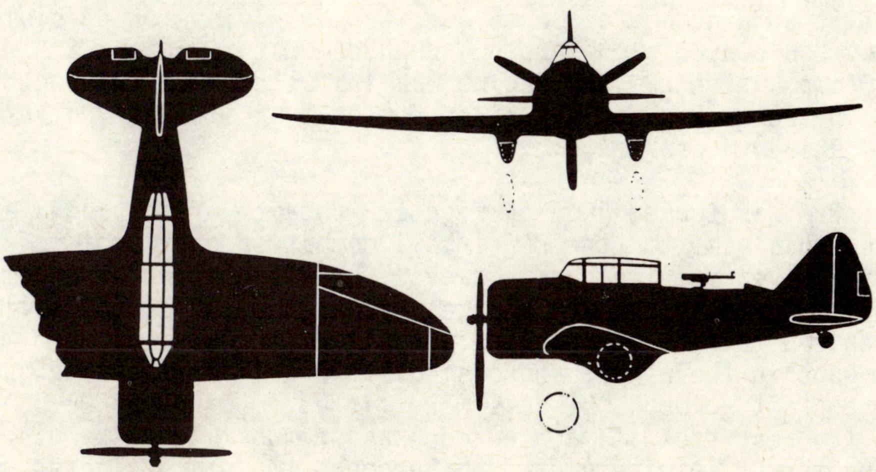

SEVERSKY 2PA

Seversky Convoy Fighter

The Seversky 2PA was all metal construction. It was designed so that flight stresses were distributed over the entire wing surface. Wing construction consisted of multiple spars and beams. No one beam carried a critical load. The 2PA was well stressed and if the aircraft took hits from machine gun bullets it would still hold together unless a vital part was hit, like the engine or oil cooler.

The 2PA had long range, because of its integral wet wing fuel tanks. The fuel capacity could be changed from any amount up to 600 gallons which provided the plane a cruising range of 3500 miles. This would equal a transcontinental flight across the United States non stop with a reserve amount of fuel. Another advantage was the safety factor. The type of construction Seversky used was inherent buoyancy and this eliminated the necessity of flotation gear, and this added safety feature would prove beneficial to foreign countries which would operate these aircraft over water.

Maj. Seversky sent the Model 2PA on a South American tour in 1939 and demonstrated it to a number of South American countries for possible sales. This Seversky 2PA played a dual role. When demonstrating the 2PA as a pursuit plane a 1000 h.p. Wright R-1820 engine was installed. If the South American country was in the market for a trainer replacement aircraft he installed a 550 h.p. P&W R-1340 "Wasp" engine.

The Model 2PA was tagged "Convoy Fighter" because of its long range. Maj. Seversky used this name for the 2PA in all company promotional work. The 2PA was armed with seven machine guns, however, provisions and arrangement of guns were optional. Normally the 2PA mounted either two .30 or .50 caliber machine guns synchronized to fire through the propeller. Provisions were made for two .50 caliber guns in the wings, and a single .30 caliber flexible gun mounted in the rear cockpit. Two bomb racks were fitted to carry a 600 pound load. Mounting of the rear cockpit machine gun was unique. Seversky dispensed with the heavy gun turret so widely used in European two place fighters. Gun turrets added hundreds of pounds of weight to the aircraft and caused unneeded drag. The Model 2PA was so designed that the entire rear cockpit was spacious, comfortable, and unobstructed for the gunner. The rear gun was mounted inside the aft fairing of the rear gunners cockpit in such a manner as to retain streamlining, light weight, and speed. Release of a small latch allowed the entire housing to revolve around the rear gun and be retracted into the fuselage. The rear gunner seat was mounted within a light weight, all metal ring so the gunner could easily shift direction. The rear gunner's role was to protect the tail of the aircraft. He could traverse his gun 260 degrees.

While demonstrating the 2PA "Convoy Fighter" in Argentina in 1939, Seversky's test pilot Frank Sinclair faced a serious problem. When the ship docked at Buenos Aires it was found that the wingspan of the Model 2PA was too great to allow its passage through the streets to the military airport located on the other side of the city. A time schedule had to be met. Rather than take the time to dismantle the aircraft, Frank Sinclair suggested he be allowed to take off with his mechanic from the 200 foot pier and fly to the Argentine Air Force field. Permission was granted from both civilian and military authorities and people from miles around came to watch. The takeoff was made without incident and the scheduled appearance of the Model 2PA was made on time.

Performance of the Seversky 2PA with a Wright R-1820 engine of 1000 h.p. based on a fuel load of 200 gallons is as follows: Wing Span 36 or 41 feet, length 25 feet 4.5 inches, height 9 feet 6 inches. Wing area 220 sq. feet or 250 sq. feet, wing loading 24.1 pounds per sq. feet. Empty weight 4034 pounds. Useful load 1918 pounds, gross weight 5952 pounds. Max. speed 300 m.p.h. at 16,500 feet, cruising speed 270 m.p.h. at 12,000 feet, landing speed 65 m.p.h. Rate of climb 3150 feet per minute. Service ceiling 30,000 feet and a range of 2600 miles with 350 gallons of fuel.

Seversky 2PA-BX "Convoy Fighter" which was demonstrated in Europe by Maj. Seversky.

JAPAN A8V-1 NAVY PURSUIT

SEVERSKY 2PA-B3

Several Seversky 2PA-B3's (A8V-1) were used by Asahi Press as fast courier aircraft.

In 1937 the Japanese Government became interested in Seversky 2PA long range "Convoy Fighter." The war in China clearly pointed out the deficiency of the Imperial Japanese Navy Air Force for long range fighter. It had an outstanding short range fighter, the Mitsubishi A5M, code name "Claude," however it did not have the range necessary to escort Japanese Navy Nakajima G4M "Nell" bombers into the interior of China.

Seversky A8V-1 was operated by the Japanese Navy as the Type S Two seat long range fighter.

JAPAN

Asahi Press A8V-1 in flight. Note civil registration number J-BAAN.

A purchasing commission was sent to the United States in 1937 to negotiate a contract for the purchase of 20 Seversky 2PA-B3 aircraft and necessary spare parts for use by the Imperial Japanese Navy. The Japanese Navy designated these aircraft as A8V-1, a two place land based fighter. The aircraft were delivered by boat in 1938 in their natural all metal finish. The aircraft were quickly assembled at Yokosuka Naval Base and were painted in the standard Japanese Navy dark green and brown camouflage topside and light gray on the bottom. They were assigned to two squadrons of Japanese Navy Air Force and were dispatched to the Japanese airbases on the coast of China. They operated from these bases for a time but all had been returned to the Japanese homeland by the time of the Pearl Harbor attack, December 7, 1941.

As one of Japan's leading newspapers, the Asahi Press was given several Seversky ASV-1 aircraft for use as fast courier planes. They flew many news missions providing up to the minute dispatches to Japan's three main islands as well as transporting its newspaper reporters to distant points and hard to reach locations in Korea, Manchuria, and China.

One Seversky A8V-1 aircraft was given to Mitsubishi Aircraft Company to disect and reveal its construction secrets.

Seversky 2PA-A amphibian for Russia featured 3 blade propeller. Aircraft was fully armed. Note 30 caliber M.G. in rear cockpit. 1st Flight was March 15,1938. Aircraft sent to Spain to fight in Spanish Civil War. Powered by Wright R-1820 of 850 h.p.

In 1938 the Russian trading commission in the United States, "Amtorg," contacted Maj. Seversky about purchasing the manufacturing rights for Seversky Aircraft, especially the Model 2PA "Convoy Fighter" for the Russian Air Force.

Russia was a very large country of vast open lands, ranges and tundra. The Russian Air Force badly needed a long range fighter to reach the far outlying air force bases under its control. If war came in 1939 Russia would be ready.

Maj. Seversky had to look to foreign sales to survive and keep his small company alive in a harsh business world. No large orders were seen for his aircraft by the Air Corps and so he had no choice but to look overseas to market his aircraft products. His selection of marketable aircraft was limited to military aircraft; he had no commercial transport designs to fall back on as did other American aircraft companies.

It is reported that Maj. Seversky signed a contract with the Russian trading commission for one million dollars; included in this contract was one Seversky 2PA-A, two place amphibian Registration No. NX1307 and one Seversky 2PA-L Registration No. R-189M, "Convoy Fighter," complete manufacturing drawings, some design tooling, and the rights to manufacture the aircraft in Russia only.

It is not known how many of Seversky designed aircraft were built in Russia or if any were manufactured at all, however in the German Air Force "Luftwaffe" aircraft recognition handbook of July 1938 under Russian aircraft, one can find photographs and 3-view drawings of the 2PA-A Amphibian and 2PA-L "Convoy Fighter," as types that could be encountered by German forces.

Russian 2PA-L was powered by Wright "Cyclone engine of 850 h.p.
Max. speed was 275 m.p.h. fully loaded. Aircraft was fully armed.

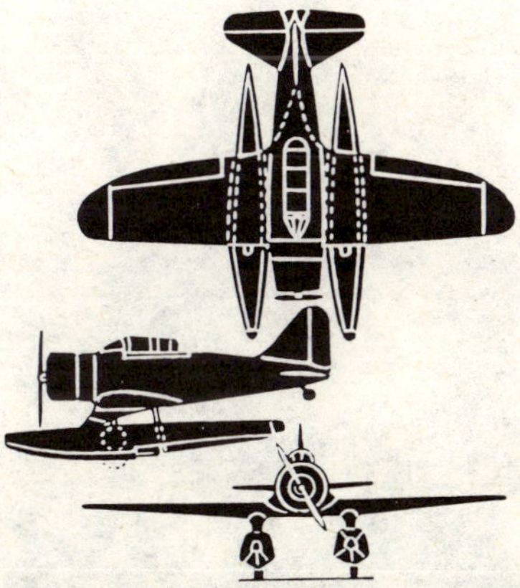

RUSSIA

Russia purchased this Seversky 2PA-A amphibian for service in Spain.
Note gun sight and exposed rear 30 caliber machine gun.

Flight tests were conducted on Seversky 2PA-L "Convoy Fighter" at factory with company test pilots and Russian observers from Amtorg Trading Corporation. Tests were conducted with full military equipment during flight evaluations.

Seversky 2PA-L had landing gear fairing attached to wing to prevent mud build up in standard Seversky wheel pant during operations from muddy dirt and grass airfields. Test pilot and Russian observer are shown here preparing to take off on test flight. U.S. Government would only permit the Wright "Cyclone" engine version of the Seversky 2PA to be exported.

SEVERSKY EP 1-106 "EUROPEAN PURSUIT & 2PA "GUARDSMAN"

Seversky EP 1=106 pursuits were known as J.9 fighters in Swedish Air Force.

In 1939 the Royal Swedish Air Force began to search for a new modern low wing fighter to replace its aging English built Gloster "Gladiators." At home the Swedish aircraft industry had nothing new to offer. Therefore, government officials were forced to look overseas. One of the current fighters that caught the eye of Swedish Air Force officials was the U.S. Army P-35 Pursuit. Alexander P. de Seversky was invited to Sweden to demonstrate his aircraft to the Swedish Air Force. Maj. Seversky arrived in 1939 and personally made all the demonstration flights before the Swedish authorities. Maj. Seversky brought with him his latest design, the Seversky EP-1-68 aircraft Reg. No. NX 2587. It was similar to the Seversky AP-7 with which Jacqueline Cochran was setting speed records in the United States. It mounted a new and powerful Pratt-Whitney R-1830 engine of 1200 h.p.

His mission was a complete success, for the Swedish government ordered 120 fighters for a little over one million dollars. Fifteen were ordered June 29, 1939, 45 were ordered October 11th, 1939 after the outbreak of World War II. A further order for 60 more aircraft was placed January 1, 1940. The first deliveries of aircraft to the Swedish Air Force began during the first half of 1940 when the first 60 aircraft were delivered. 60 of these aircraft were Seversky EP 1-106 single place fighters similar to the U.S. Air Corps P-35. The other two aircraft were Seversky 2PA-204A "Convoy Fighters." The EP 1-106 was designated J-9 in Swedish Air Force service and the Seversky 2PA-204A's were known as B-6 two place light bomber aircraft.

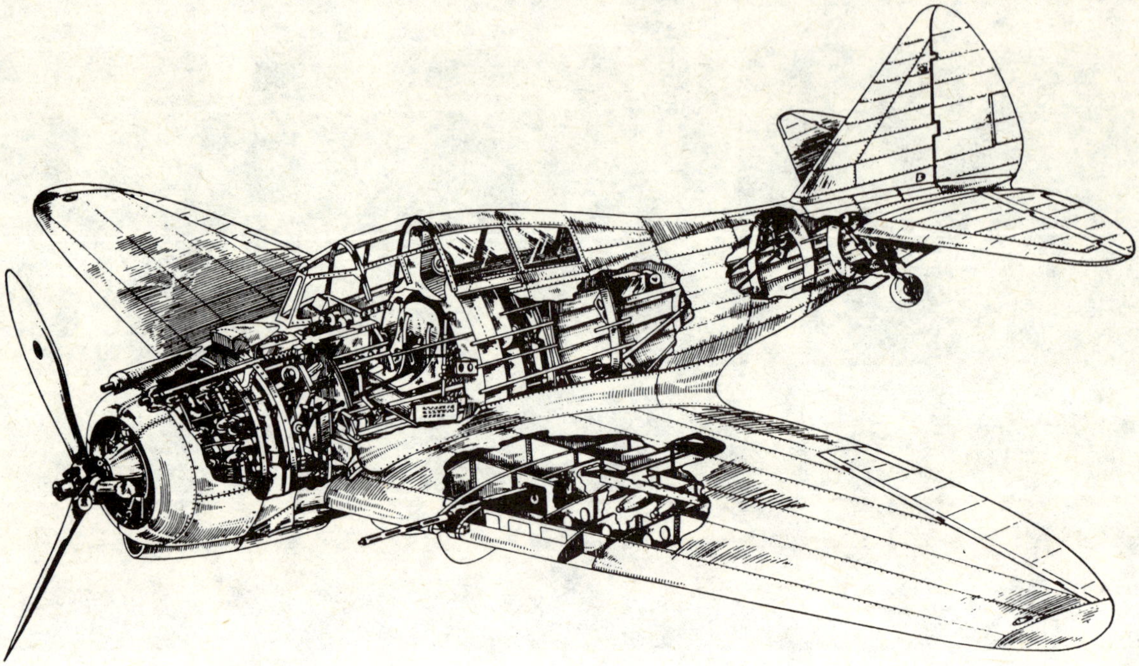

Sweden ordered 120 Seversky EP 1-106 pursuits but only 60 were delivered

The 60 Seversky EP 1=106 aircraft seized by the U.S. Government were identical to the Swedish J.9 aircraft with the exception of the prop spinner.

Seversky pursuits were used to defend Swedish air space during W.W.II.

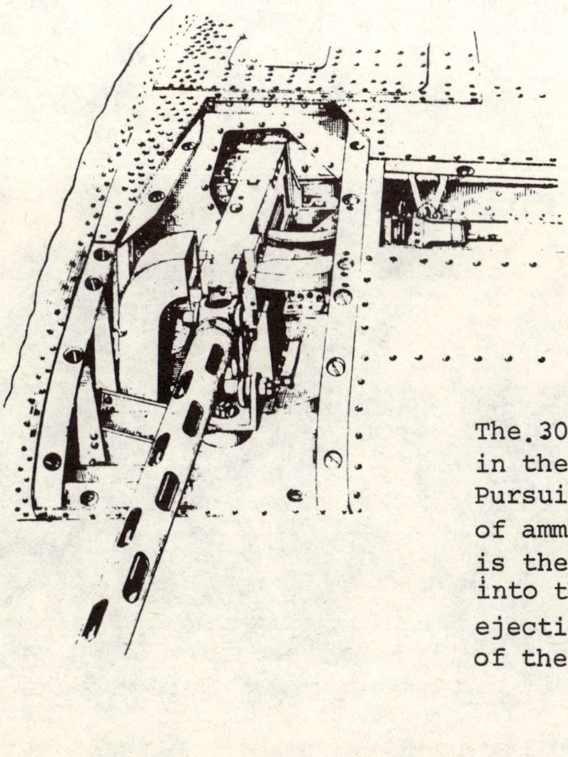

The.30 caliber machine gun as installed in the wing of the Seversky EP 1-106 Pursuit was capable of firing 1000 rounds of ammunition. The container at the right is the ammunition box feeding cartridges into the gun through the chute. The link ejection chute is located just to the left of the gun.

Mechanic makes final check out of Seversky Pursuit just prior to final test flight at Seversky factory on Long Island, New York. Aircraft was then dismantled, packed into a box, and shipped to Sweden.

The remaining aircraft on order were taken over by the U.S. Government. Following the seizure of 60 EP 1-106 fighters by the United States, the Swedish Government ordered 60 Reggiane Re. 2000 "Falco" fighters in 1940 from Italy. This look alike fighter was a direct copy of the Seversky P-35. The Reggiane Re. 2000 was an all metal, single place fighter powered by an 18 cylinder Piaggio P XI R.C. 40 radial engine of 870 h.p. It carried two 12.7 machine guns in the cowl. It had a wingspan of 36 feet 1 inch, length 25 feet 11 inches and had a gross weight of 5,700 pounds. It had a cruising speed of 280 m.p.h. and a max. speed of 330 m.p.h.

Swedish Air Force pilots liked to fly the Seversky Pursuit, and Swedish Air Force mechanics stated the aircraft were easier to maintain than any other Swedish Air Force fighter including the Italian built Reggianne Re.2000.

When the Germans persuaded the government of Finland to fight Russia, Sweden was surrounded by forces of conquest. Sweden remained neutral during these hazardous times and the Seversky EP 1-106 pursuits played a very important part in defense at a time when war surrounded this peaceful country.

The Seversky EP 1-106 remained in service with the air force as a front line fighter from 1940 to 1947. Later they were used as advanced trainers. A number were modified into photo-reccon aircraft and equipped with two aerial cameras mounted in the baggage compartment. By 1953 ten EP 1-106's were still flying. The chief of the Swedish Air Staff, Gen. G.A. Westring and Gen. B.M. Jacobsson were each flying Seversky pursuits on inspection tours throughout the country as their personal aircraft, long after the type was retired from active fighter status.

The Seversky aircraft in Sweden gave a good account of itself after more than thirteen years of service. The aircraft was rated excellent, particularly from the maintenance point of view. During its service years with the Swedish Air Force, only eight crashes occurred and none of these crashes were suspected to have been caused by material defects. The average flying time per aircraft amounted to between 1000 to 1500 hours total flying time. It was an outstanding record of rugged aircraft.

Sweden ordered 50 Seversky 2PA-204A "Guardsman" but only two were delivered.

COLUMBIA

SEVERSKY 3M-WW

In 1935 the Columbian government had a requirement for a high speed amphibian aircraft, capable of operating from both land and water. The search for such an aircraft focused upon the Seversky 3M. After conferring with Maj. Seversky, the Columbian government placed an order for three Seversky 3M-WW aircraft. This aircraft was a three place twin float amphibian aircraft in which the landing gear retracted within the floats.

This was the first production contract which Maj. Seversky received for his new aircraft company. The three amphibian aircraft for Columbia were based on the first Seversky SEV-3 of 1933. The three Seversky 3M-WW mounted the Wright Whirlwind engine of 450 h.p. By December 1936 all three Seversky 3M-WW amphibian aircraft had been delivered for service with the Columbian government where they were used as reconnaissance and patrol aircraft.

A follow-on order was placed by the Columbian government and three more Seversky 3M-WW aircraft were purchased, bringing the total order to six aircraft. This was the largest number of Seversky amphibian aircraft ever ordered by a single country.

All metal SEV-3M-WW amphibian was of rugged construction

Twelve Seversky P-35A Pursuits were shipped to Ecuador by the United States under a mutual aid program in 1942. These aircraft were part of a shipment of aircraft seized from Sweden.

In January 1941, the government of Ecuador invited the United States government to send a mission to Ecuador to aid and resupply the Ecuadorian Air Force with more modern aircraft.

Upon arrival, the United States Air Mission found the first item of importance was to train the pilots in modern flying techniques. Several Fairchild PT-19's and North American NA-16's were sent down to help teach Ecuadorian pilots at Punta Salina.

The first fighter squadron formed was supplied with a dozen surplus Seversky P-35A's seized by the United Stated from Sweden in 1940. These P-35A aircraft were used by the Ecuadorian Air Force throughout World War II until replaced by more modern P-47 "Thunderbolt" fighters in 1947. All the P-35A's were unfortunately scrapped.

MEXICO - EXPORT

In 1936 the Seversky Aircraft Corporation sold its first aircraft, the SEV-3M to a Mexican Army officer. This aircraft was powered by newly installed Wright R-1820 of 850 h.p.

It is believed that this aircraft was later resold to Republican forces fighting in Spain during the Spanish Civil War. A recent attempt was made to locate this aircraft but to date no leads have turned up, thus it must be presumed that the aircraft was destroyed in the Spanish conflict.

DOMINICAN REPUBLIC - EXPORT

A total of at least two Seversky aircraft were purchased by the government of Dominican Republic. The Seversky AP-9, Registration No. NX-2598 was sold in summer 1940. This was the aircraft that Jacqueline Cochran had flown and set a new women's 2000 km. speed record of 322 m.p.h. on April 6, 1940. The aircraft had many new innovations and parts from other Seversky aircraft. It had an inward retracting landing gear from the Seversky AP-8. The fuselage came from an early Seversky, however when it was completed in mid-1938, it was 30 m.p.h. faster than the AP-7 had been. This aircraft had also been flight tested at Wright Field in the pursuit design competition held February-March 1939, but it was not considered for an Army contract and was passed over. It had flush riveting throughout the wing, a new airfoil and a sharp leading edge to the wing which Jacqueline Cockran complained about as giving her some flight problems when she was flying the aircraft.

The Seversky EP-1-68 U.S. Civil Registration No. NX-2587 was built in 1938. It was similar to the Seversky AP-7, and was powered by a large new Pratt-Whitney R-1830 engine of 1200 h.p. when delivered to Dominican Republic. This aircraft had been used by Maj. Seversky in Sweden as his demonstrator aircraft, hence the designation EP "European Pursuit." It was with this aircraft that he received an order totaling one million U.S. dollars.

The aircraft registry for the government of El Salvador shows that a Seversky 2PA aircraft was registered in El Salvador in 1951 to Mr. A. Baldocchi. This aircraft was purchased by Mr. Baldocchi for the purpose of testing various electronic test apparatus in the early 1950's.

Seversky 2PA shown here without wheel pants and in El Salvador Registration YS-114.

V SEVERSKY P-35A's GO TO WAR

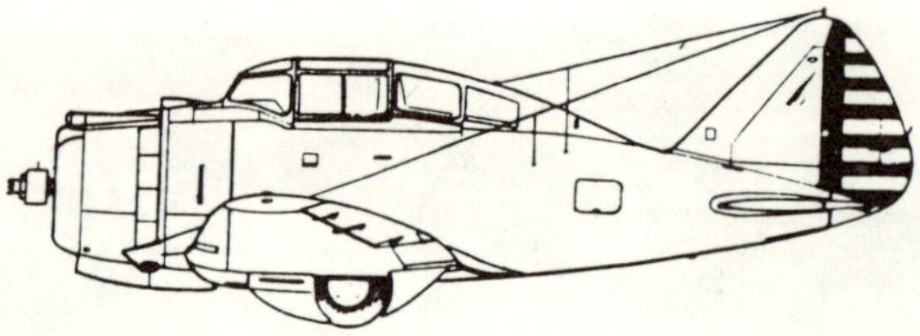

Sixty EP-1-106 fighters destined for Sweden were seized in 1940 by the U.S. Government. These aircraft were designated P-35A's and 48 were sent out for defense of the Philippine Islands in 1940. They arrived by boat at Manila and were trucked to nearby Nichols Field for assembly. They were test flown and issued to squadrons.

The P-35A's bore Swedish instrumentation and placards in the cockpit area. Each morning a class was conducted to teach Air Corps pilots the Swedish language to fly an American built pursuit in a Spanish speaking country. Irony had played its part.

Dual transition training was done in North American A-27 attack type aircraft which had been placed in emergency service to requalify newly arrived pilots into the flying game after the long sea voyage from the states. The A-27 was similar to the AT-6 trainer that the pilots had flown in stateside flying schools, however it mounted a powerful 850 h.p. Wright R-1820 engine which gave it a max. speed of 250 m.p.h.

Prior to the arrival of the P-35A's the various pursuit squadrons in the Philippines had been flying antique P-12E and P-26A "PeaShooter" type pursuit aircraft. The P-12E's were given to the Philippine Government for pursuit trainers and later when the first Filipino Pursuit Squadron was formed the P-26A's were also handed over. These same P-26's were later used in combat by the Philippine Air Force pilots against the invading Japanese.

P-35A's of 17th Pursuit Sq. at Nichols Field 1941.

AT-12 "GUARDSMAN"

Fifty AT-12 all metal, two place export fighters entered service with the Army Air Corps in 1940 in an unusual manner. These aircraft were originally ordered and paid for by the Swedish Government as the Model 2PA "Guardsman," a two place fighter-dive bomber. On October 10, 1940 these 50 aircraft were seized by the United States War Department for the Army Air Corps along with 60 EP-1-106 Export Fighters also ordered by Sweden. The Swedish Government protested vigorously but U.S. Secretary of State Cordell Hull stated that the emergency in Europe justified the seizure.

Model 2PA-204A in flight over New York countryside. Two of these aircraft were delivered to the Swedish Air Force before the U.S. Government seized the other 50 aircraft.

The AT-12 was a two place export development of the P-35 A It could serve as a two place, long range fighter, or be fitted as a dive bomber. It had an increased wing span and carried a larger fuel load. It mounted two .30 caliber Browning machine guns in the cowl, firing through the propeller and a .30 caliber machine gun mounted in a flexible gun ring in the rear cockpit. In addition a 1300 pound bomb load could be carried nearly 1000 miles.

The Air Corps had no operational requirement for an aircraft of this type, however they did acquire them for service. The AT-12 was more of a combat type aircraft than a trainer. The dive bombing rack was removed. It is reported that they were used for a time as advanced trainers, however, the AT-6 "Texan" filled this vital role gallantly from 1940 to 1946. Later the AT-12 filled another vital need. Requests were received by base airfield commanders at various Air Force installations scattered throughout the United States for a fast personal aircraft transport to attend Air Force conferences, etc. and to keep current in fast combat type aircraft. And so a number of AT-12 aircraft were dispatched to serve the needs of various base commanders. One AT-12 spent its entire Air Force career at Lowry Field, Denver, Colorado. One AT-12 was sent to Wright Field for test work. It was used as a test vehicle to record airflow distribution over various surfaces of the aircraft. It was later fitted with a special dorsal fin at Wright Field.

Two AT-12 "Guardsman" Aircraft survived W.W. II. The other 48 aircraft were either written off in accidents or cut up for scrap following the war.

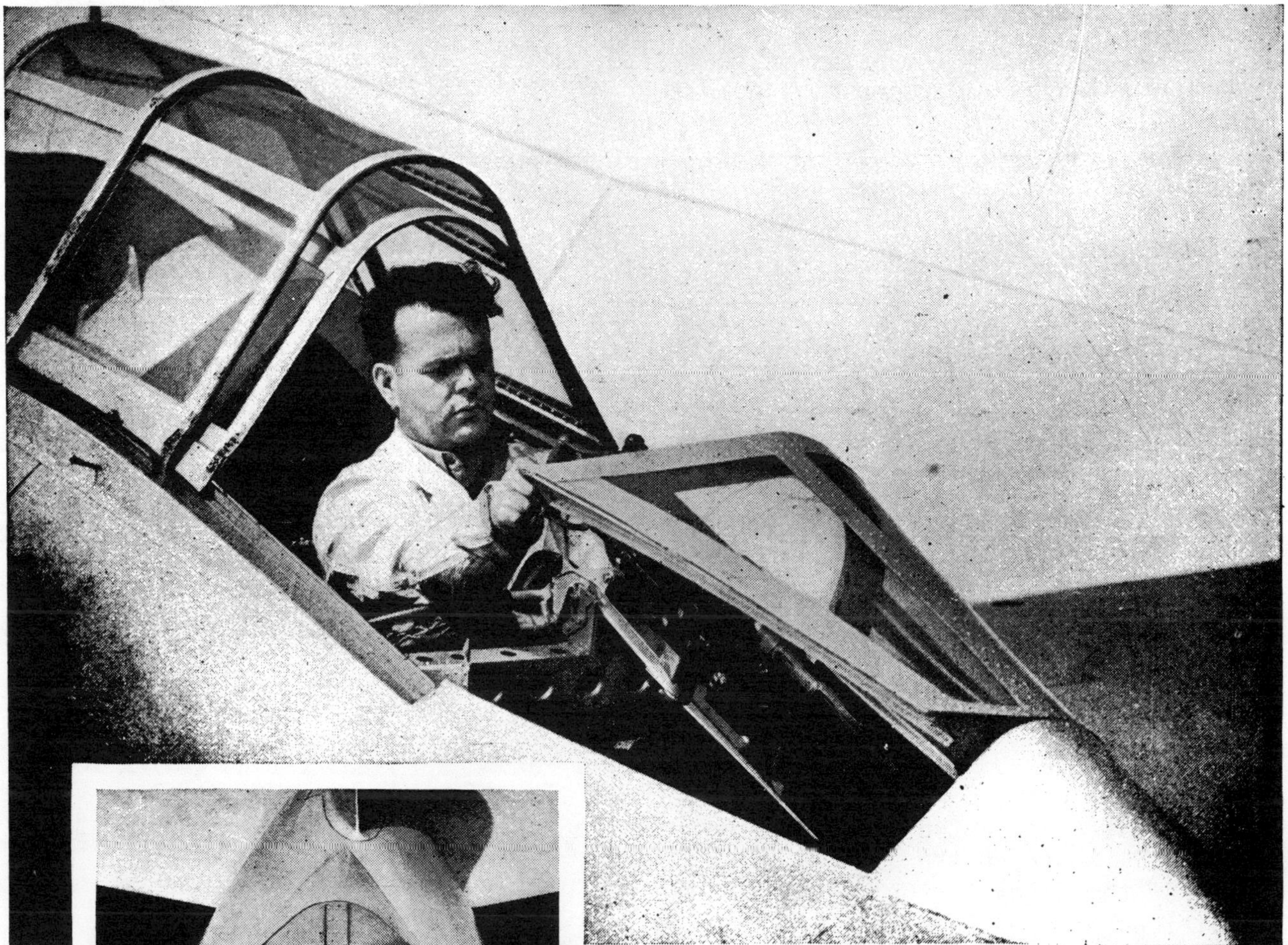

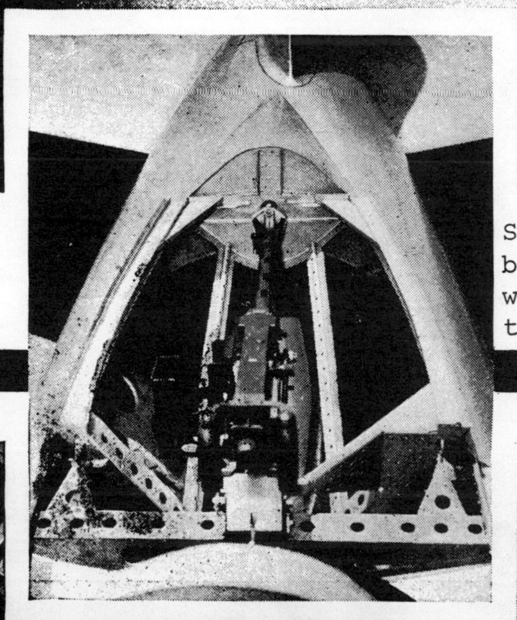

Seversky 2PA rear gunner position. Turtle back canopy rolls over and under out of the way. Seat revolves to enable gunner to defend tail position. Gun is 30 caliber Browning.

guns to the rear!

P-43 "LANCER"

The P-43 "Lancer" was a development of the Seversky P-35. In early 1939 Seversky offered the Air Corps two design proposals: the XP-41, a medium altitude fighter with inward retracting landing gear and fitted with a turbo supercharger; and a new design known as the Seversky AP-4, which mounted a turbo supercharger in the belly of the fuselage.

The original AP-4 featured a tight fitting cowl and a large spinner in an attempt to eliminate drag, but cooling problems arose and a normal radial engine cowl was fitted.

Thirteen service test models of the AP-4 were ordered May 12, 1939 by the Air Corps and were designated YP-43's. These aircraft were named "Lancer," and were powered by Pratt-Whitney R-1830-35 engine of 1200 h.p. It had a maximum speed of 350 m.p.h. at 20,000 feet. It mounted two .50 caliber machine guns on the cowl and two .30 caliber guns in the wings. The first YP-43 was delivered in September 1940. The last YP-43 was delivered in April, 1941. Some changes were made in the YP-43 from the original AP-4 design. The cockpit was lowered and the rear fuselage decking was raised. The transparent plexiglass area aft of the cockpit was greatly reduced, and the tail wheel strut was lengthened.

On September 9, 1939, an order was placed for 80 XP-44 aircraft. This order was later increased to 827 additional XP-44 aircraft. Combat reports from Europe later indicated that the XP-44 design was obsolete. On September 13, 1940, all XP-44 contracts were cancelled in favor of the new P-47 design. In view of the delay in tooling up for the new P-47B the Air Corps ordered an additional 54 P-43A "Lancer" aircraft with R-1830-35 engines to keep the production line open.

Nostalgic and colorful in aviation history are the U.S. fighters that took wing between the two world wars. Theirs was a glamorous era that bridged plane development from stick and wire to the supersonic all metal jet fighters of today. Because the type designation of these early fighters was the letter "P" they became known affectionately as "The PeaShooters" to the pilots who flew them.

P-35 design influence can be clearly seen in YP-43 "Lancer".

The order for 80 XP-44's was changed to 80 P-43A's with R-1830-45 engines. The first P-43 was delivered in May 1943 and were virtually identical to P-43. Delivery of the first of 80 P-43A's was made September 1941.

On June 28, 1941, 125 P-43A-1's were ordered under Lend Lease by China for the Chinese Air Force. The P-43A-1 had the R-1830-57 engine, mounted four .50 caliber machine guns, more armor, and self sealing fuel tanks. Production of P-43A-1 "Lancer" aircraft was completed in March 1942. Some 108 were delivered to China via bases in India. One was flown over the top of Mount Everest by Col. Robert L. Scott, in 1943.

The Air Corps modified 150 "Lancers" to photo-reccon aircraft and these were designated P-43B's. Some were supplied to the Royal Australian Air Force in 1942. Two P-43's were modified into P-43C's and the P-43D and E models were proposed reccon versions.

One of 13 YP-43 "Lancer" high altitude pursuits of 1939.

P-43 "Lancer" was immediate predecessor to famous W.W.II P-47 "Thunderbolt" No P-43's exist today. The last surviving P-43 was cut up for scrap at Freeman Field, Seymour, Indiana in the early 1950's.

YP-43 was single place, all metal fighter with turbo supercharger for high altitude combat.

Chinese Air Force P-43A flies alongside captured Japanese A6M-2 Zero Fighter.

Chinese Air Force P-43A "Lancer" takes off for mission from China air base.

XP-44 "WARRIOR"

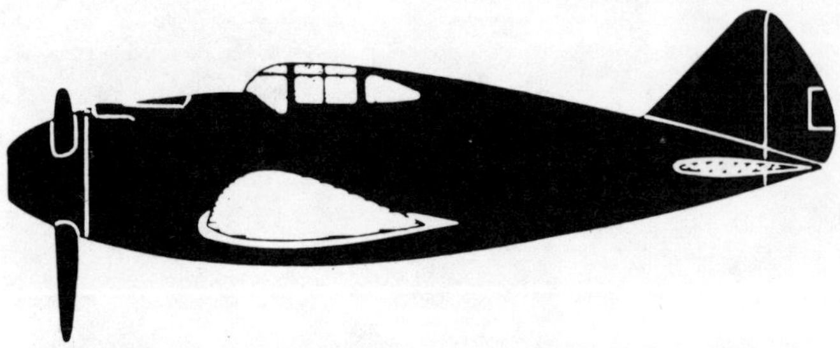

Alexander Kartveli designed and had worked on a new proposed fighter design for some time during the development of the XP-41, and XP-43 "Lancer." Many long hours of work had gone into this new design and a great deal of research had gone into the XP-44. Over one million dollars had been spent in testing wind tunnel models, design layout, engineering drawings, etc.

The XP-44-1 was an all metal, low wing monoplane design. It was a traditional Seversky airplane with designer Kartveli's fine touch. It was flush rivet construction throughout, and it was streamlined from nose to tail. Up front it mounted a huge spinner and a very tight fitting cowl to reduce parasitic drag. A Curtiss Electric 3 bladed propeller was to have been fitted. A new Pratt-Whitney R-2180-1 engine of 1400 h.p. was to have been mounted in the nose. It had a wingspan of 36 feet, length of 28 feet 4 inches.

When this new P&W R-2180 engine was not forthcoming, Kartveli designed the XP-44-2. This design mounted the most powerful radial engine developed up to that time, the P&W R-2800-1 18 cylinder engine of 2000 h.p.

Alex Kartveli held great hopes for the XP-44-2 design. He and several company engineers were invited to attend a fighter conference at Wright Field. Here he was given the opportunity to meet and talk with a number of visiting pilots who had been in combat in the skies over Europe. Kartveli came away convinced that he had to go back to the drawing board and start over, for the air war over Europe was rapidly changing. If America was to have the best possible radial engine, high altitude fighter he would begin anew. He knew that if XP-44-2 was built, it would be so overloaded it would not meet the design specifications he had outlined, and so the XP-44 "Warrior" design was dropped after a million dollars had been expended. This tough decision paid great dividends later when Kartveli designed the XP-47B.

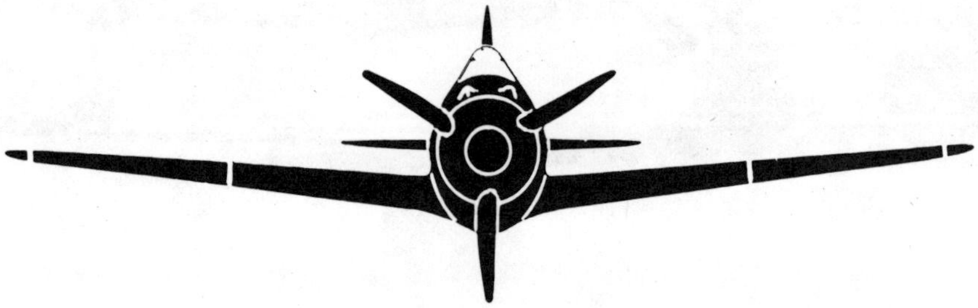

"THUNDERBOLT"

"Thunderbolt" Fighters: XP-47B, P-47B and YP-47B. Note different canopy enclosures.

After Alexander Kartveli dropped his XP-44 design he felt an obligation to begin anew and design a newer and more modern fighter. He came up with a new design, the AP-10 designated XP-47 and XP-47A by the Army. These aircraft both mounted an Allison V-1710 liquid cooled engine. This fighter was to have a maximum speed of 400 m.p.h. and climb to 15,000 feet in 3½ minutes.

Two prototypes were ordered in January 1940. One XP-47 and one XP-47A. Both aircraft were identical but the XP-47A had armament and other military equipment. Both aircraft were cancelled when military combat reports from Europe showed these two aircraft would not answer the needs of the next generation fighter.

Undismayed Kartveli returned to his drafting board and came up with a new and larger fighter design and mounting for the new powerful Pratt-Whitney R-2800 engine originally proposed for the XP-44-2 design. This new radial engine, coupled with a turbo supercharger offered exceptional performance at high altitude. Kartveli also built in long range fuel tanks, making it one of the longest ranging fighters of the period. It was soundly designed, strongly built and had the heaviest armament of any fighter in the world: eight Browning .50 caliber machine guns mounted in the wings.

The XP-47B S/N 40-3051 was completed in early 1941. It made its first flight May 6, 1941 by Republic test pilot Lowry Brabham. Upon landing after the first test flight and taxiing up near company officials watching his flight, he got out and shouted "We hit the jackpot - what a ship."

XP-47B "Thunderbolt" in flight December, 1941.

The number two airplane of a long line of "Thunderbolts" the YP-47B poses for photograph. Note pre-war marking.

P-35 & P-43 geneology is clearly evident here in design of XP-47B. Aircraft had O.D. top side, and gray bottom. U.S. Army was in black.

Eight gun "Thunderbolt" was heaviest armed U.S. fighter of W.W.II. to go into production. Some 15,683 were built during the war.

History was made that day so very long ago. The XP-47B was nearly twice the size of other fighters. A four bladed Curtiss Electric propeller was fitted and a maximum speed of 400 m.p.h. was achieved. Considerable testing of the XP-47B prototype was conducted throughout the summer of 1941. The first production orders were received in the fall of 1941 for $58,000,000.00 and first deliveries to the Air Force were made in January 1942. Thus the P-47 "Thunderbolt" was born. The outstanding combat achievements in the skilled hands of American and Allied pilots during the war placed it above any fighter the enemy put into the air. Some 15,660 "Thunderbolt" Fighters were built, making it the largest production fighter in American history. It had the lowest mission combat loss rate of any American fighter in Europe. When the "Great Book of Immoral Aircraft" is written the P-47 "Thunderbolt" will be one of the top selections.

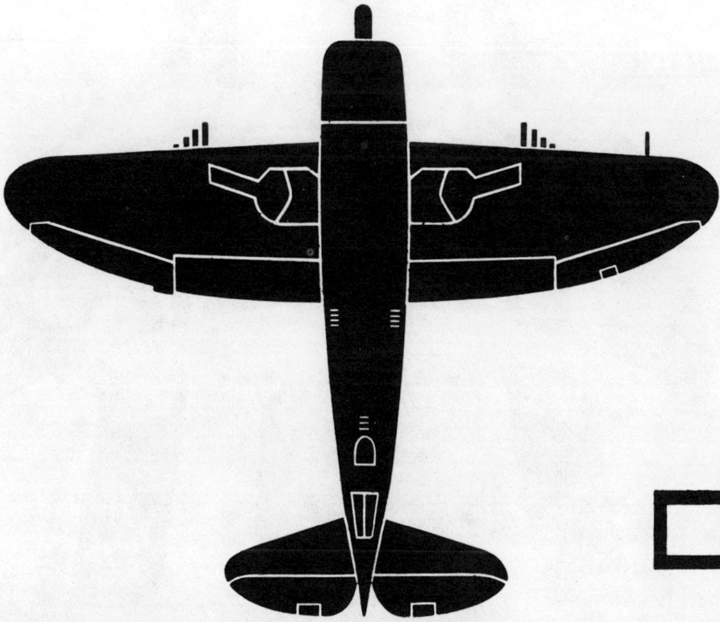

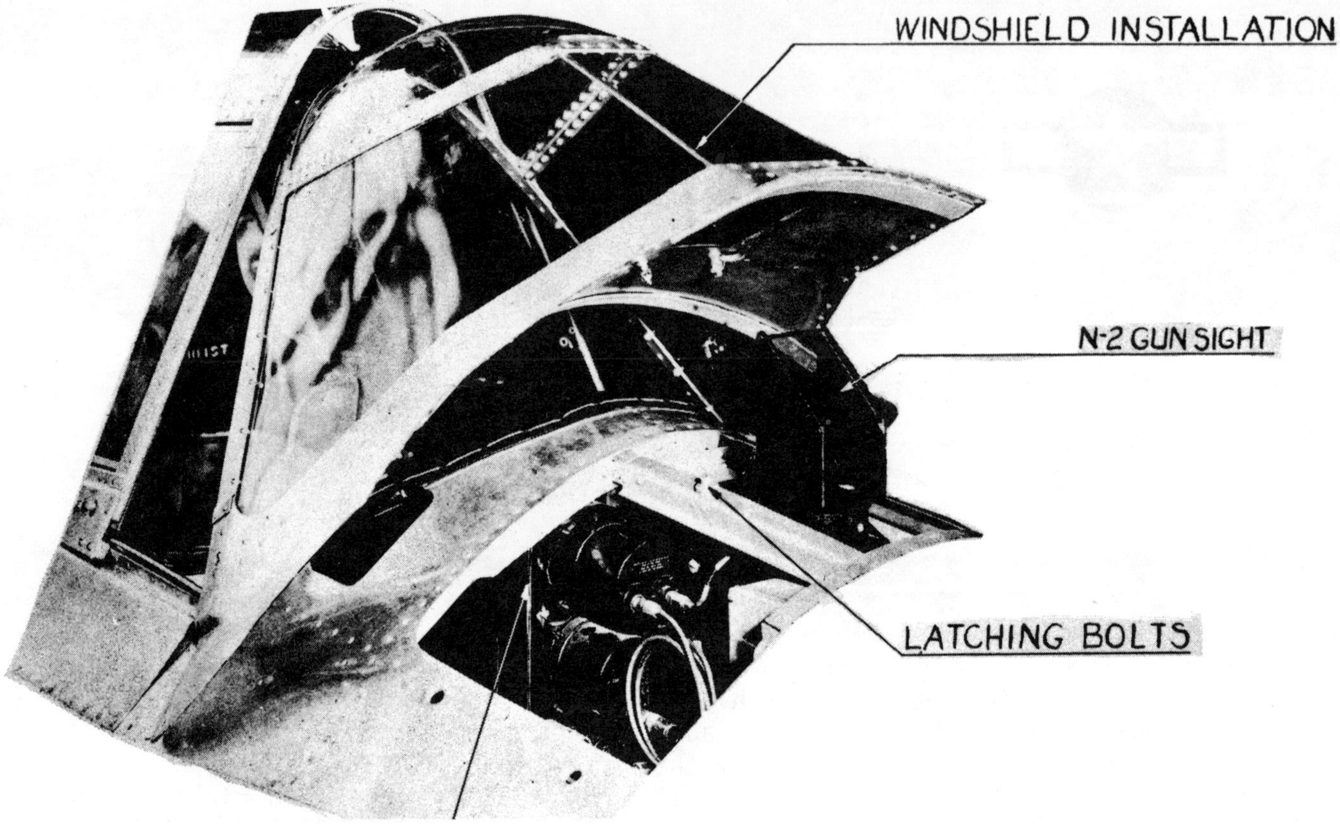

Seversky P-35 Windshield and N-2 Gunsight Installation

Browning 50 cal. and 30 cal. Machine Gun Installation As Installed in P-35 Pursuit.

PRATT & WHITNEY R-1830 "WASP" ENGINE

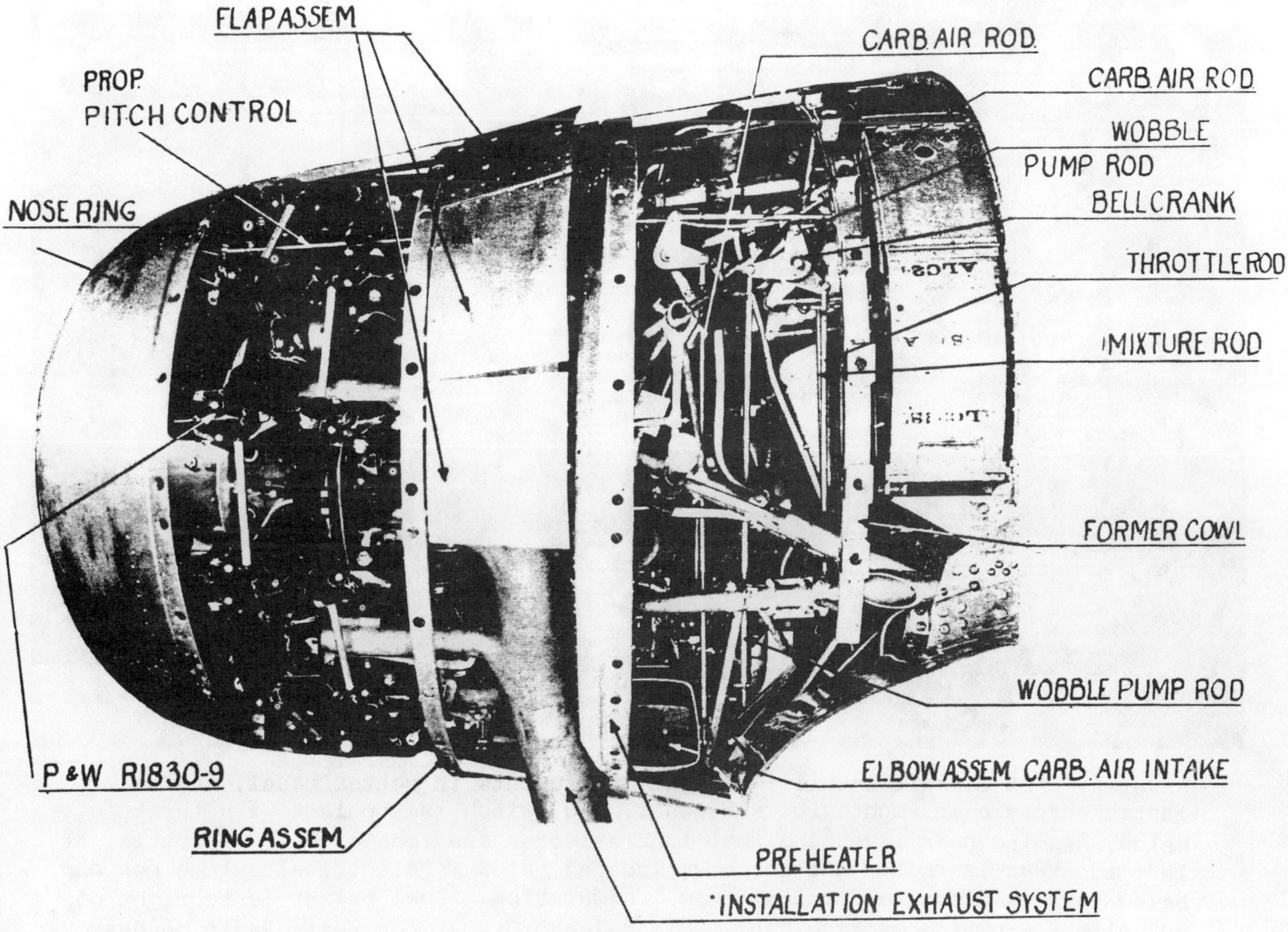

Left side P-35 engine details with cowling removed

Seversky P-35 Cockpit Details: Flight instruments in center panel, engine instruments are in right side sub-panel. Mag.switch is top left with throttle below. Landing gear lever and fuel tank selector are located below throttle. At left top rear is radio junction box, located below is electrical switch box and below this is aleron and rudder trim. RIGHT SIDE: Fuel primer is to right of 30 caliber Browning machine gun. Radio selector is at top rear. Radio headset box is located below and pilots mike is located between it and sub-panel. Hot and cold air selector and oxygen control is located at right rear.

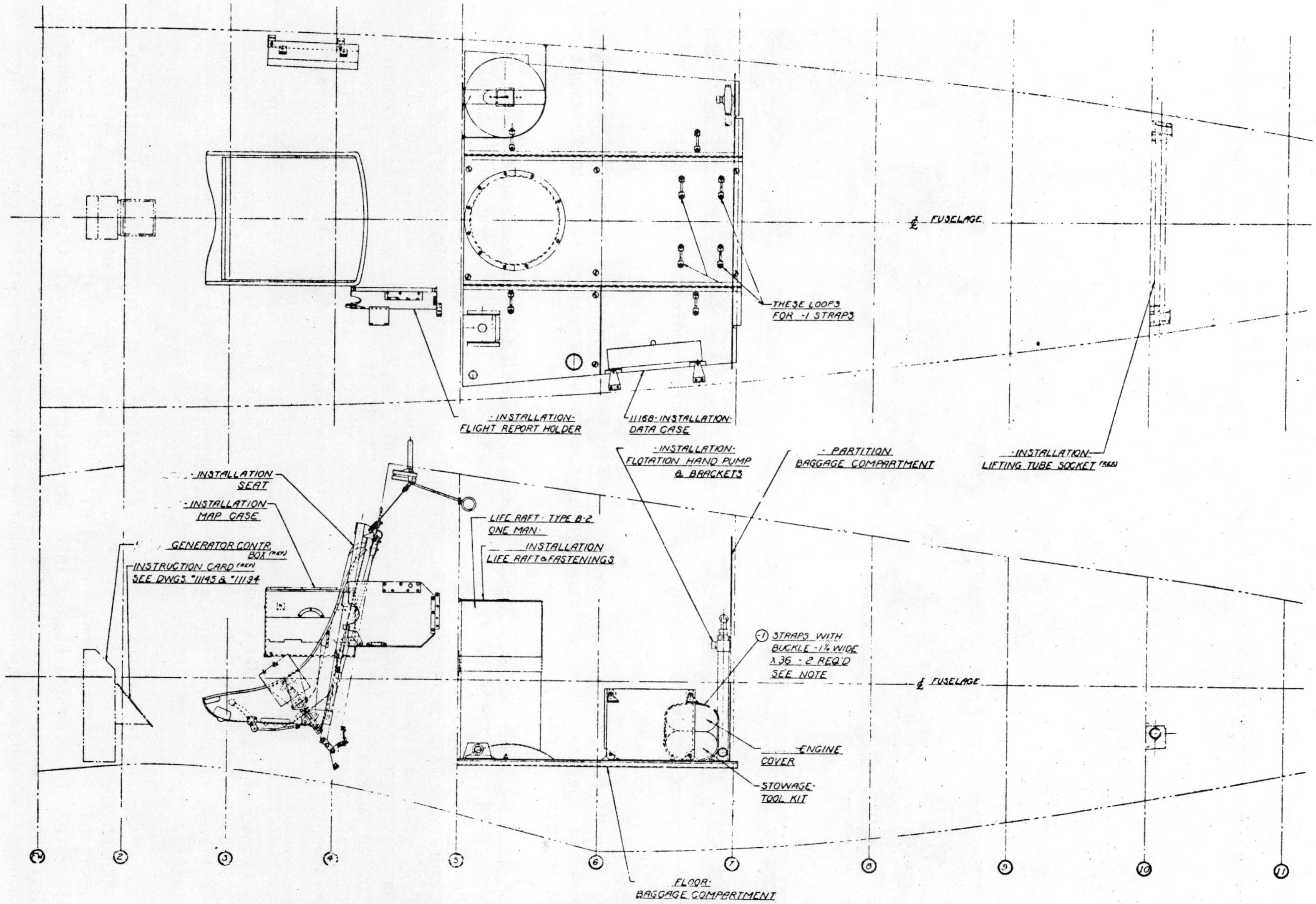

SEVERSKY P-35 FUSELAGE EQUIPMENT DIAGRAM AND PILOTS SEAT INSTALLATION

Restoration completed,P-35 proudly awaits dedication ceremony.

One of the latest additions to the Air Force Museum at Dayton, Ohio is an original Seversky P-35 which was acquired from Charles P. Doyle of Rosemont, Minnesota.

This P-35 was one of a number given to vocational trade schools during World War II to help train mechanics. It somehow survived the scrap man's breaker bar and axe and Mr. Doyle acquired the aircraft in the middle fifties from the trade school. The Air Force Museum traded for the P-35 aircraft in 1974. This P-35 once served with the 1st Pursuit Group at Selfridge Field, Michigan. It was assigned to the 94th Pursuit Squadron. The P-35 was officially presented to the museum in a celebration on June 25, 1974. The aircraft was restored to display condition by maintenance personnel of the 133rd Tactical Airlift Wing, Minnesota Air National Guard, under the command of B/General John R. Dolny.

This P-35 is in excellent condition. It is painted in the colors and markings of the squadron commander of the 94th Pursuit Squadron. During the roll out celebration Maj. Alexander P. de Seversky attended. This grand old gentleman flew in all the way from New York to attend the ceremony on June 2, 1974.

During the inspection period, Maj. Seversky was cautioned about slipping when on the wing of the P-35. Maj. Seversky's comments were "Don't worry, I designed this bird for one legged pilots," and with that he lifted his artificial leg over the side and hopped into the cockpit. This was the last P-35 function Maj. Seversky attended for he passed away a few months later. The grand old man is gone but one of his remarkable fighter designs will be available for inspection for generations to come.

The grand old man himself. Maj.Alexander P.de Seversky at the P-35 dedication spoke of the development of the aircraft and the love he had for this plane.

Sgt. Gene Andreotti supervised restoration of Air Force Museum's P-35.

Maj. Seversky mounts P-35 Pursuit for first time in 36 years.

Maj.General Dolny assists Maj. Seversky onto P-35 wing.

Maj. Seversky waves to dedication crowd from P-35.

Recalling memories of days gone by.

SEVERSKY P-35A

P-35A after restoration by Air Force Museum personel. Note markings of Lt.Buzz Wagner 17th Pursuit Sq.

In 1971 the Swedish Air Force donated one of its Seversky EP 1-106 (P-35A) fighters to the Air Force Museum as an expression of the close friendship which has existed between our two countries.

This aircraft was originally one of 60 Seversky EP 1-106 ordered by the Swedish Government in 1939, shipped by boat to Norway and then shipped by rail train to Sweden.

The Air Force Museum staff and restoration personnel have done a terrific job of restoring this aircraft which was badly pitted on its aluminum skin. After being stored under ground for so many years the salt caused the skin to corrode, however machine polishing has done a remarkable job in bringing back the original metal luster.

This aircraft has been repainted in the colors and markings of America's first Air Force "Ace"of World War II, Lt. Boyde "Buzz" Wagner when he was commanding officer of the 17th Pursuit Squadron, Nichols Field 1941.

While in service with the Swedish Air Force this aircraft was known as the type J-9 fighter and this aircraft bore the number 26 painted on the rudder. It was flown at one time by the commander in chief of the Royal Swedish Air Force.

Sole surviving Seversky EP1=106pursuit of Swedish Air Force Museum Collection.

Today there remains in Sweden one Seversky J-9 type fighter S/N 2134. It is based at Malmalatt. It was one of 60 ordered by the Swedish Government in 1939. It was delivered on May 15, 1940. It was finally retired from flying March 31, 1949 after having flown some 820 hours.

It was one of two J-9 fighters preserved by the Swedish Air Force. They had all given a good account of their suitability in Swedish service. Today the J-9 fighter is held in storage as part of the Royal Swedish Air Force museum collection.

Ex-Swedish Air Force Seversky EP1-106upon arrival at the Air Force Museum

SEVERSKY MODEL 2PA-204 (AT-12)

Seversky 2PA in 17th Pursuit Sq. markings.

This plane is on display at Planes of Fame aviation museum, Cal Aero Field Chino Airport, California. This aircraft was originally ordered by the Swedish Government and was seized and taken over by the War Department and given to the Air Corps in 1940.

During World War II it served at Lowry Field, Denver, Colorado as the base commander's aircraft. After the war it was declared surplus by the War Assets Administration and sold. It was later registered out of the country. Later it was returned to the United States and now resides in the museum.

Down memory lane: P-51D "Mustang" and Seversky 2PA.

PASSING PARADE OF PURSUITS: Boeing P-12E & P-26A, Seversky 2PA, P-38 & P-40 at "Planes of Fame" Museum, Chino Airport, California

Seversky 2PA has pleasing lines.

SEVERSKY SEV-3 AMPHIBIAN

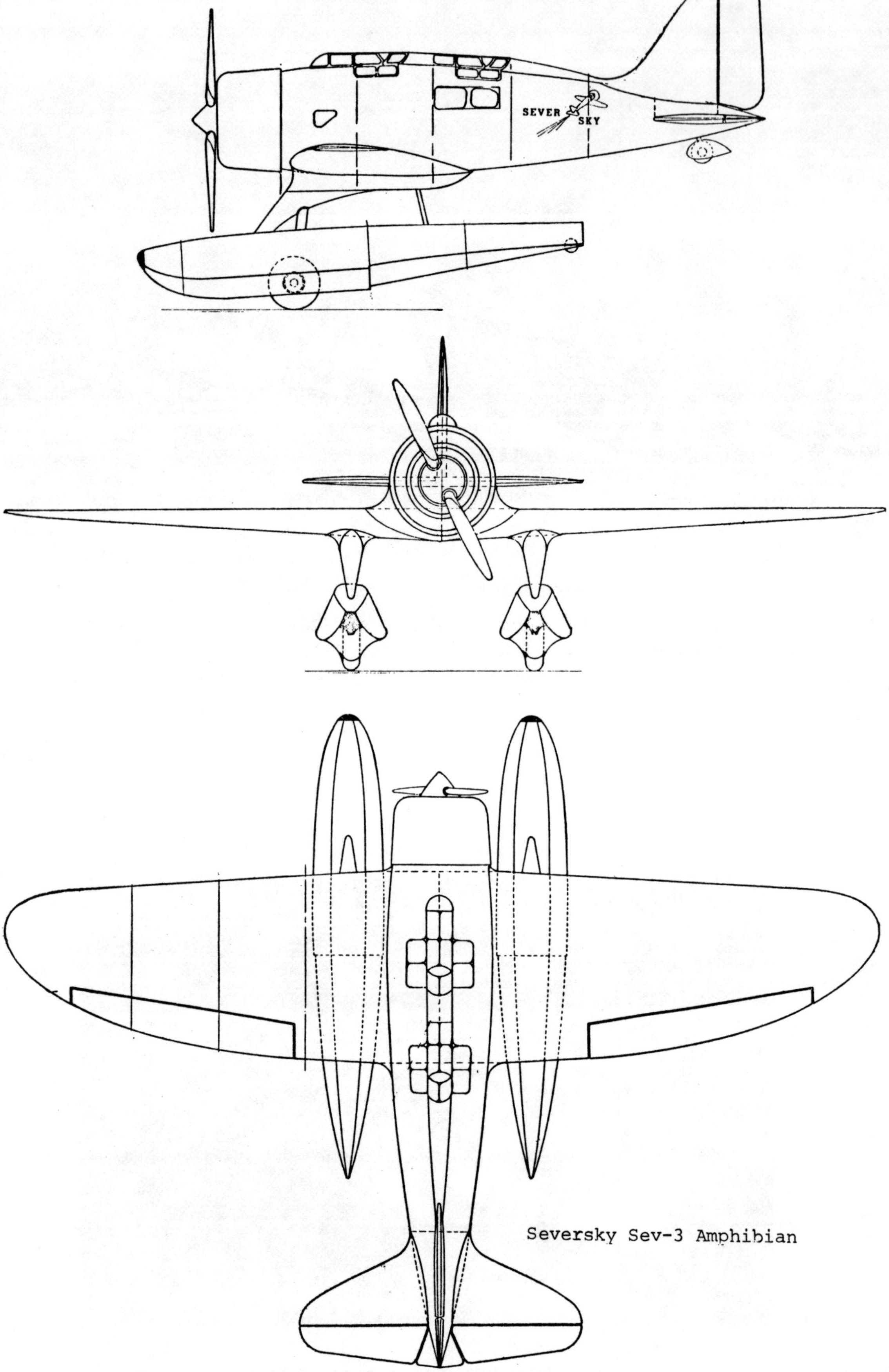

Seversky Sev-3 Amphibian

SEVERSKY BT-8 BASIC TRAINER

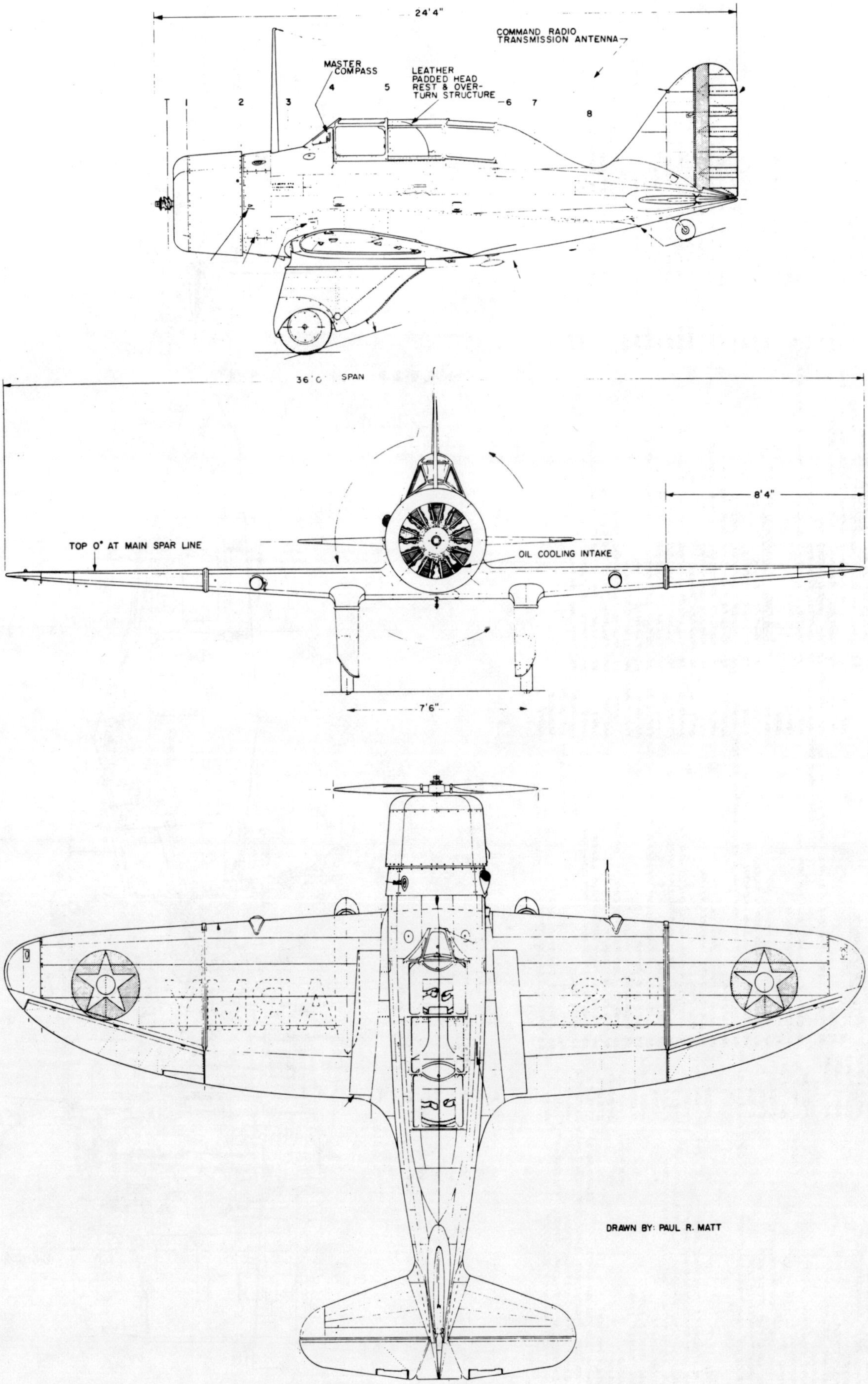

INDEX

	Item	Spec. or Dwg. No.
1.	BATTERY (10303)	TYPE C-5
	RADIO EQUIPMENT	10037
2.	JUNCTION BOX	10405
3.	TUNING CONTROL - RECEIVER	MC-125
4.	CONTROL BOX - RECEIVER	BC-( )-231
5.	CONTROL BOX - TRANSMITTER	BC-( )-232
6.	RECEIVER	BC-( )-229
7.	TRANSMITTER	BC-( )-230
8.	ANTENNA RELAY	BC-( )-198
9.	ANTENNA	11310
10.	REMOTE CONTROL UNIT	MC-135
11.	MICROPHONE (11384)	TYPE T-17
12.	DYNAMOTOR	BC-( )-83
13.	DUAL COIL UNIT FOR RECEIVER	C-155A
14.	SPARE COIL FOR RECEIVER	C-171
15.	COIL FOR TRANSMITTER	C-216
16.	SPARE COILS FOR TRANSMITTER	C-217, C-218, C-219 & C-220
	ARMAMENT	10039
17.	.30 CAL. AIRCRAFT MACHINE GUN	M-2 OR 1918, M-1
18.	.50 CAL. AIRCRAFT MACHINE GUN	M-2
19.	.30 CAL. AMMUNITION BOX & AMMUNITION	11219
20.	.50 CAL. AMMUNITION BOX & AMMUNITION	11134
21.	EJECTION CHUTE	11004
22.	.30 CAL. BLAST TUBE	11305-1 & -2
23.	.50 CAL. BLAST TUBE	11305-3
24.	SOLENOID (34B4186)	TYPE E-2
25.	PLUNGER	34B5133
26.	TRIGGER MOTOR	07G023
27.	TRUNNION BOLT & BRACKET (32D3819)	TYPE A-3
28.	POST ASSEM. - MOUNTING BRACKET (32D3823)	TYPE A-3
29.	GUN SIGHT N-2 INSTALLATION	10820
30.	GUN CAMERA G-4 INSTALLATION (ALTERNATE)	11157
31.	AMMUNITION ROUNDS COUNTER	11053

	Item	Spec. or Dwg. No.
	FLARE EQUIPMENT	10038
32.	PARACHUTE FLARES	TYPE M-8
33.	FLARE RACKS (11012)	TYPE A-3
34.	CABLES - FLARE RELEASE	10038
35.	HANDLES - FLARE RELEASE	TYPE A-3
	POWER PLANT	
36.	FUEL	104 GALS.
37.	OIL	9.5 GALS.
38.	PROPELLER - 10'-6" DIA. - 3 BLADE - CONTR. PITCH - CONSTANT SP'D	HUB-3E50-209 BLADE-6157A
39.	ENGINE - P.&W. - 850 HP. AT 2450 R.P.M.	R-1830-9
40.	EXHAUST MANIFOLD	10023
41.	PREHEATER	10781
42.	INDUCTION SYSTEM	10026
43.	CONTROL - CARB. AIR INTAKE	11520
44.	COCKPIT HEATING & VENTILATING SYSTEM	11689
45.	CONTROL - COCKPIT HEATING & VENTILATING SYSTEM	11690
46.	FIREWALL	10121
47.	RING COWL - ENGINE	10018
48.	OIL SYSTEM INSTALLATION	11521
49.	OIL COOLER	B6-28141
50.	INSTALLATION - OIL DILUTION COCK	11429
51.	FUEL SYSTEM INSTALLATION	10023
52.	FUEL GAGE	~
53.	FUEL PUMP (INCLUDING RELIEF & BY-PASS VALVE)	TYPE C-7A
54.	WOBBLE PUMP TYPE D-2	AN4009
55.	GASOLINE STRAINER	TYPE C-2A
56.	SHAFT - FUEL PUMP REMOTE DRIVE	31-2174
57.	COUPLING - SHAFT	32D-3381
58.	WOBBLE PUMP OPERATING LEVER	11363
59.	DIAL ASSEM. - FUEL COCK CONTROL	11466
60.	FUEL COCK TYPE F-2	AN4003
61.	PRIMER PARKER SPEC.	28136
62.	ENGINE CONTROL UNIT	36D1633
63.	PROPELLER CONTROL	10048

	Item	Spec. or Dwg. No.
	POWER PLANT (CONT'D)	
64.	MAGNETO SWITCH CONTROL	10494
65.	STARTER SWITCH & FOOT CONTROL	10344
66.	CONTROL BOX - GENERATOR	10240
	LANDING GEAR & TAIL WHEEL	
67.	LANDING GEAR ASSEM. - 27" DIA. WHEELS HYDRAULIC BRAKE & ELECTRIC CONTROL	10016
68.	CONTROL - RETRACTABLE LANDING GEAR	10049
69.	TAIL WHEEL ASSEMBLY	10071
70.	CONTROL CABLES - TAIL WHEEL	10030
71.	CONTROL UNIT - L.G. & FLAP POWER ECLIPSE	F-55358
	SURFACE CONTROLS	10027
72.	CONTROL STICK	10032
73.	RUDDER PEDALS	10030
74.	RUDDER CONTROL CABLES	10030
75.	RUDDER TRIM TAB CONTROL	10034
76.	RUDDER TRIM TAB CONTROL CABLES	10034
77.	RUDDER TRIM TAB	
78.	RUDDER HINGE	10266
79.	ELEVATOR CONTROL INSTALLATION	10029
80.	ELEVATOR CONTROL RODS	10029
81.	ELEVATOR TRIM TAB CONTROL	10034
82.	ELEVATOR TRIM TAB CONTROL CABLES	10034
83.	ELEVATOR HINGE	10266
84.	FLAP CONTROL	10033
85.	AILERON CONTROL LEVER	10028
86.	AILERON TRIM TAB CONTROL	10034
87.	AILERON TRIM TAB CONTROL CABLES	10034
88.	BRAKE CONTROL	10031
89.	PARKING BRAKE	10030

CABIN

90. WINDSHIELD
91. CABIN - SLIDING SECTION
92. CRASH PROTECTOR
93. TURTLEDECK
94. INSTRUMENT BOARD - MAIN
95. INSTRUMENT BOARD - SIDE
96. BAGGAGE COMPARTMENT
97. HEAD REST

SEATS & ACCESSORIES

98. SEAT - PILOT'S
99. SAFETY BELT
100. CUSHION - LIFE PRESERVER

MISCELLANEOUS EQUIPMENT

101. OXYGEN CYLINDER — TYPE B-1
102. OXYGEN REGULATOR — TYPE A-6
103. FLIGHT REPORT HOLDER
104. MAP CASE
105. DATA CASE
106. INSTRUCTION CARD
107. FLOTATION HAND PUMP
108. LIFE RAFT - ONE MAN — TYPE B-2
109. WARNING UNIT - FLAP & L.G.
110. WARNING UNIT - FUEL SWITCH-OVER
111. TAIL LIFT TUBE
112. GYRO INSTRUMENT CONTROL
113. STOWAGE FOR TOOL KIT
114. STOWAGE FOR ENGINE COVER

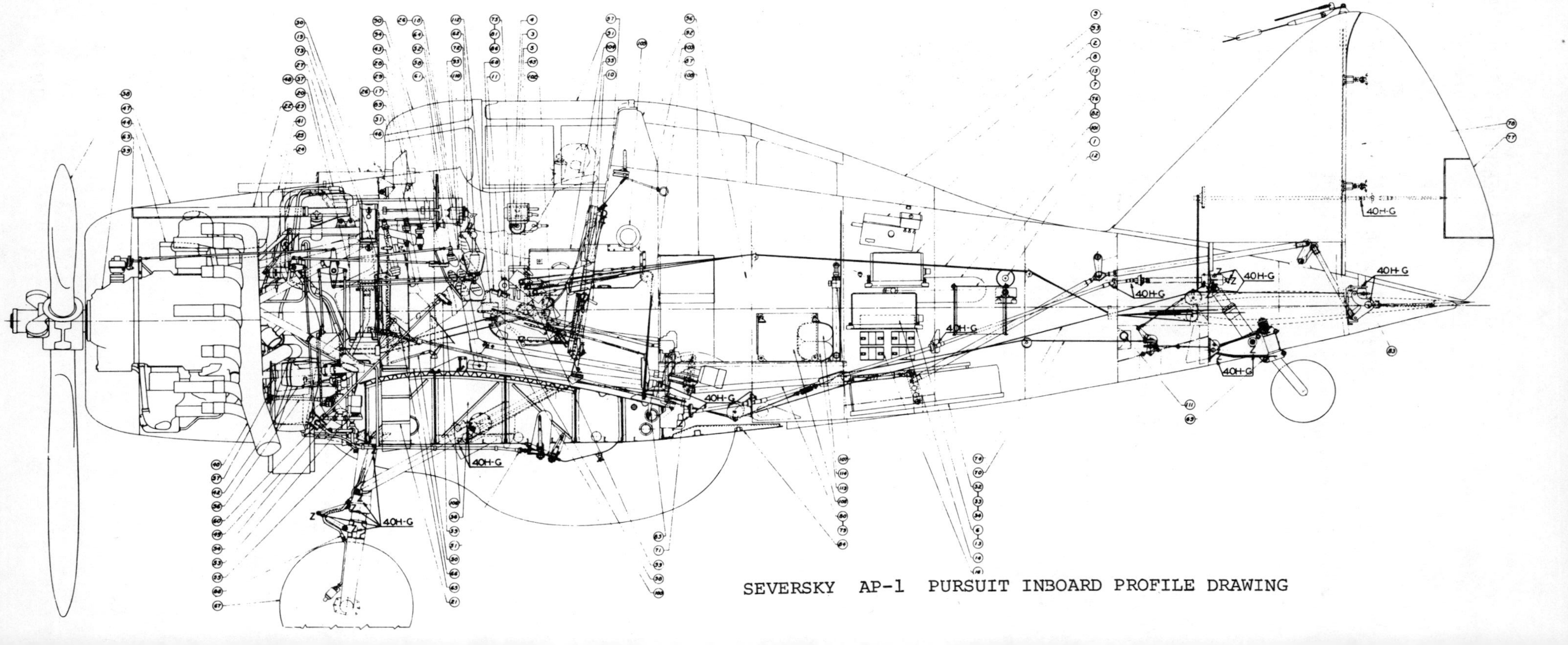

SEVERSKY AP-1 PURSUIT INBOARD PROFILE DRAWING

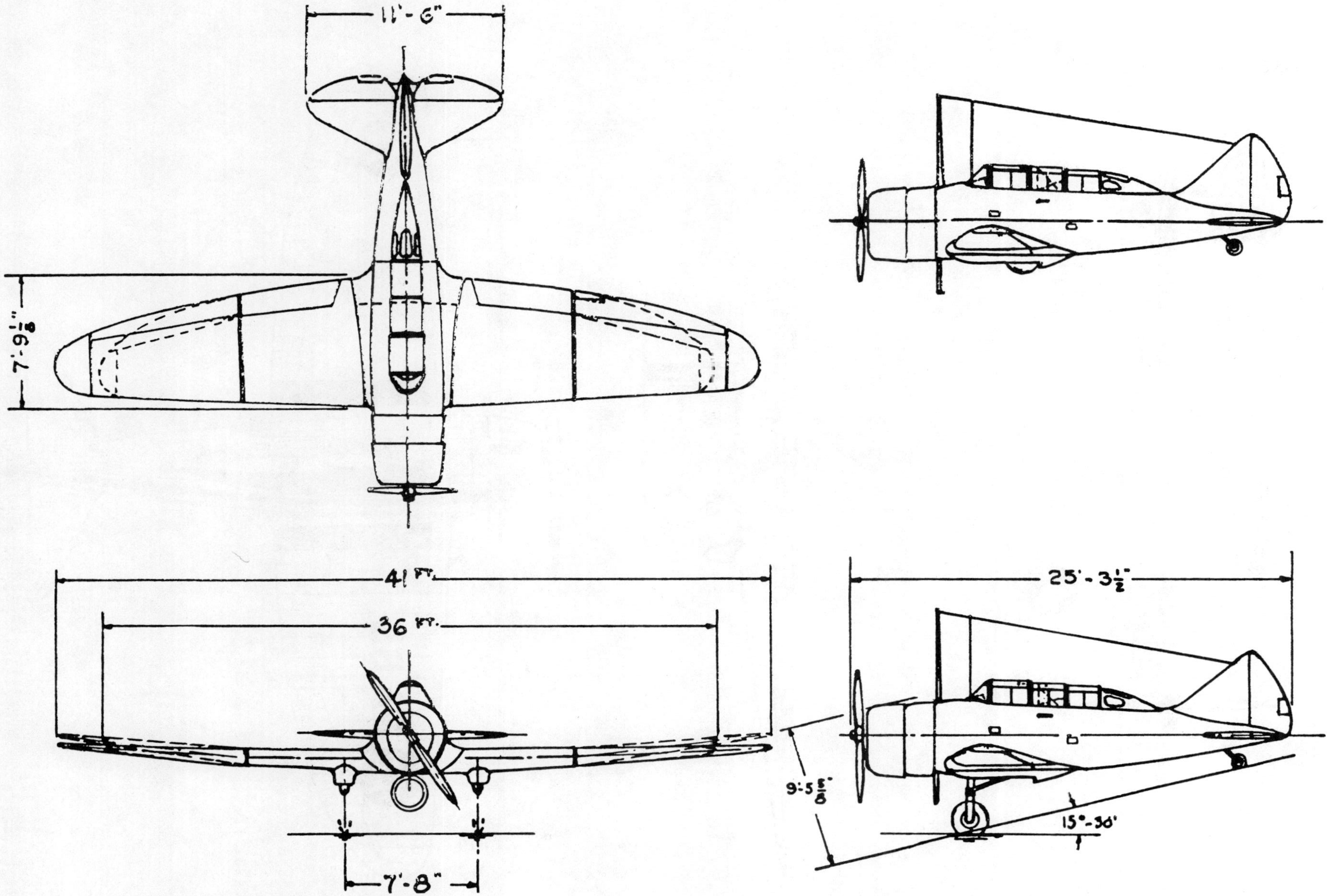

Seversky X-BT Showing Large And Small Wing Panels With And Without Dihedral.

SEVERSKY P-35 PURSUIT

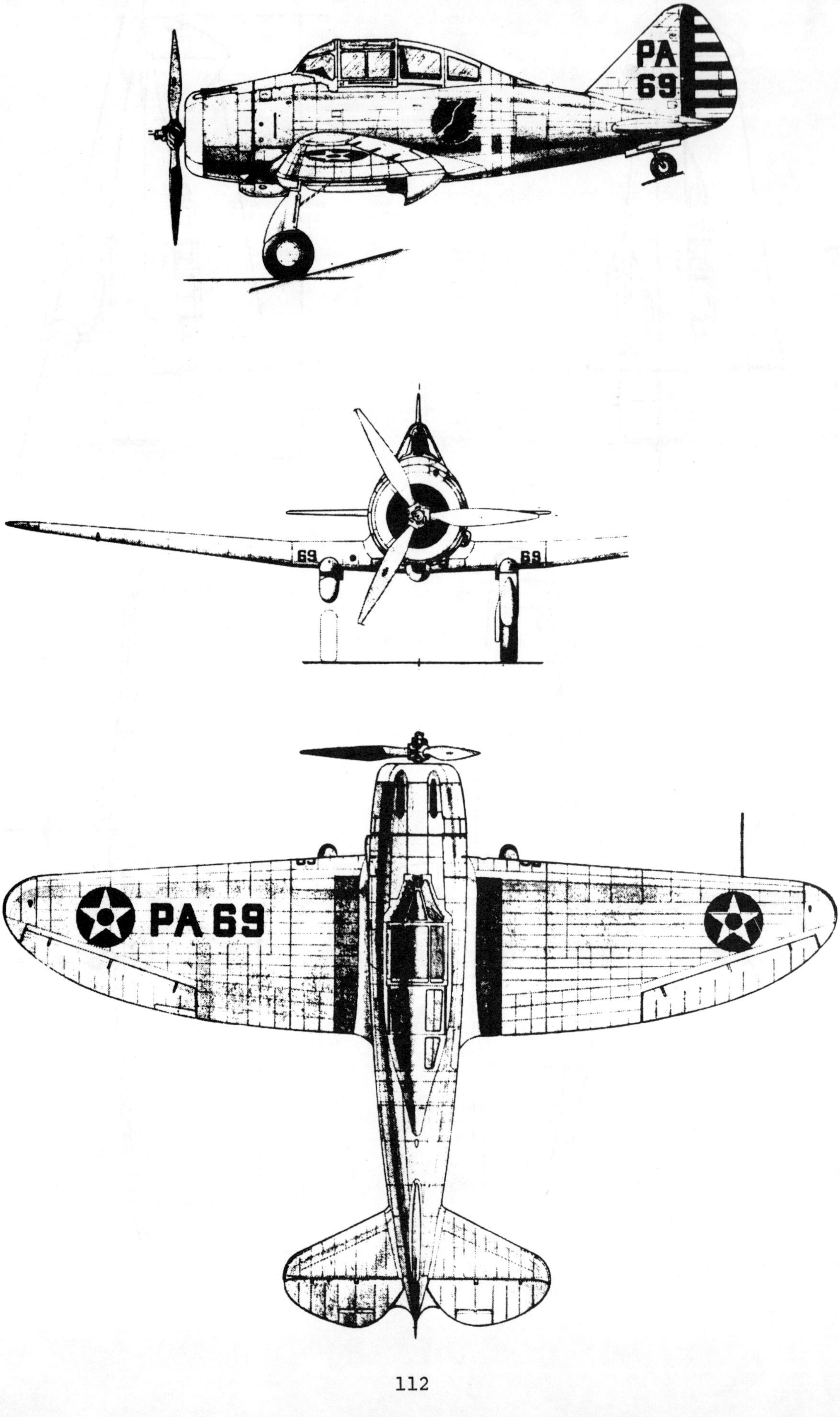

SEVERSKY P-35 A PURSUIT

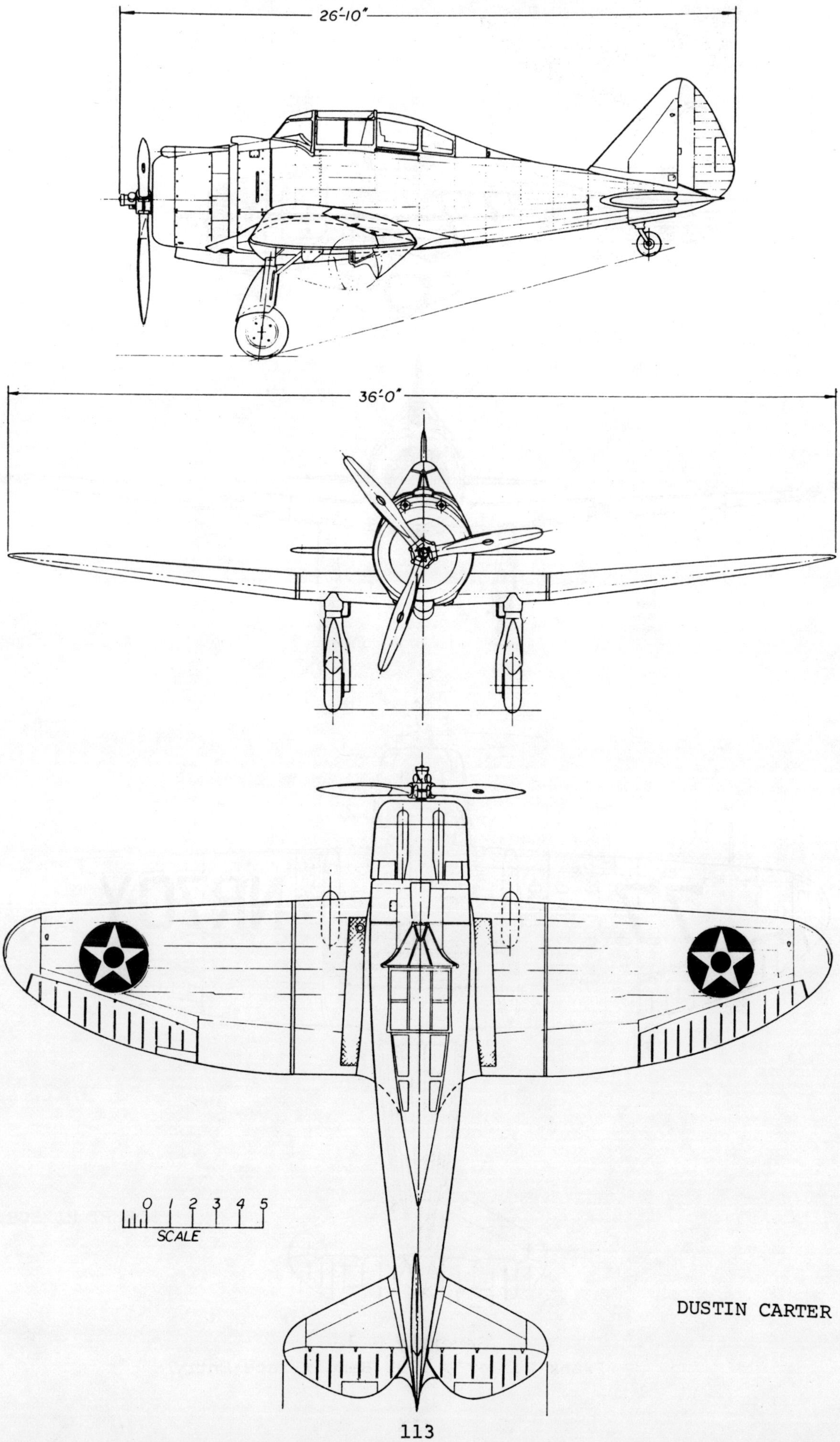

DUSTIN CARTER

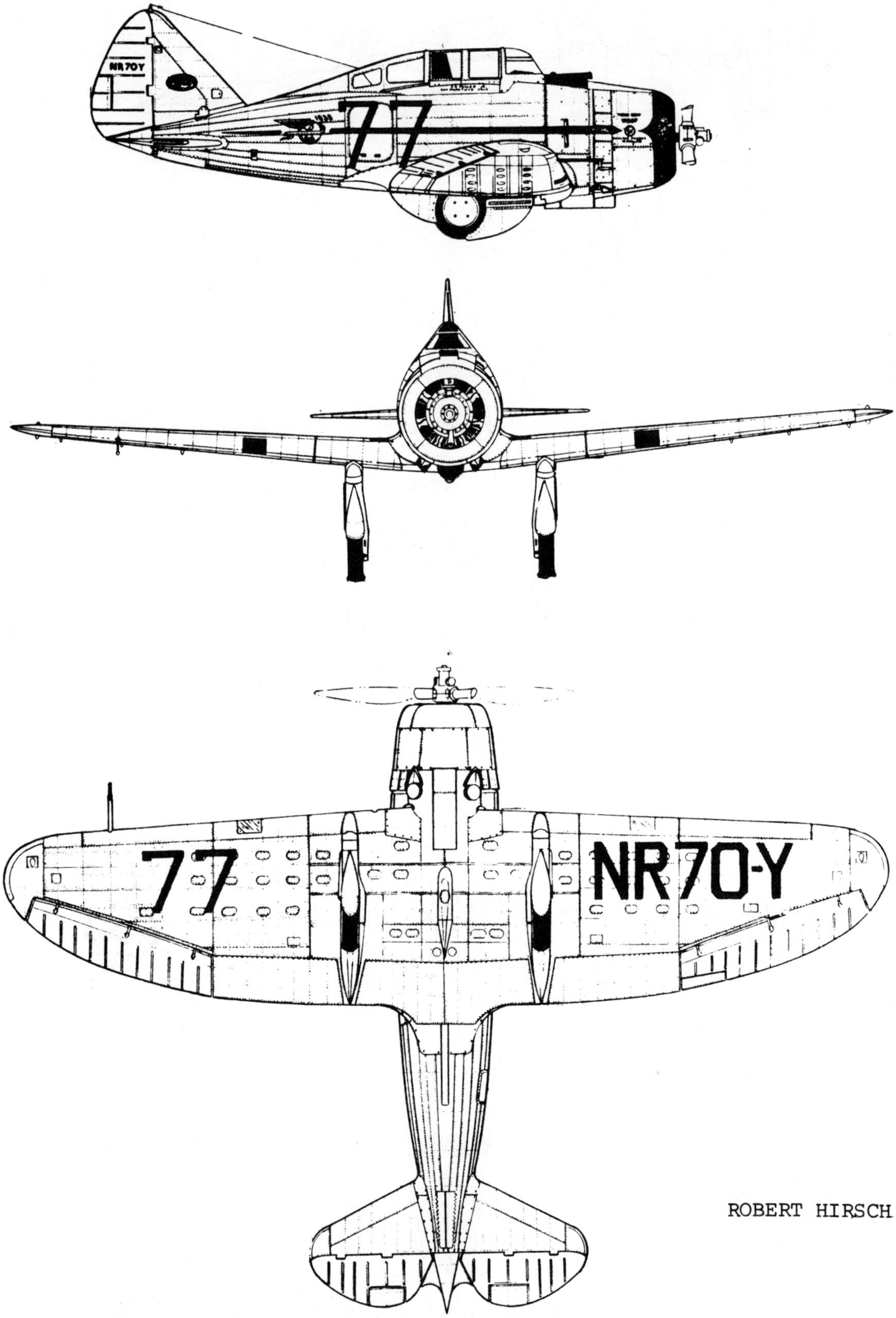

ROBERT HIRSCH

Seversky S 2
Frank Fuller's 1938 Bendix Race Entry

BENDIX RACE WINNER

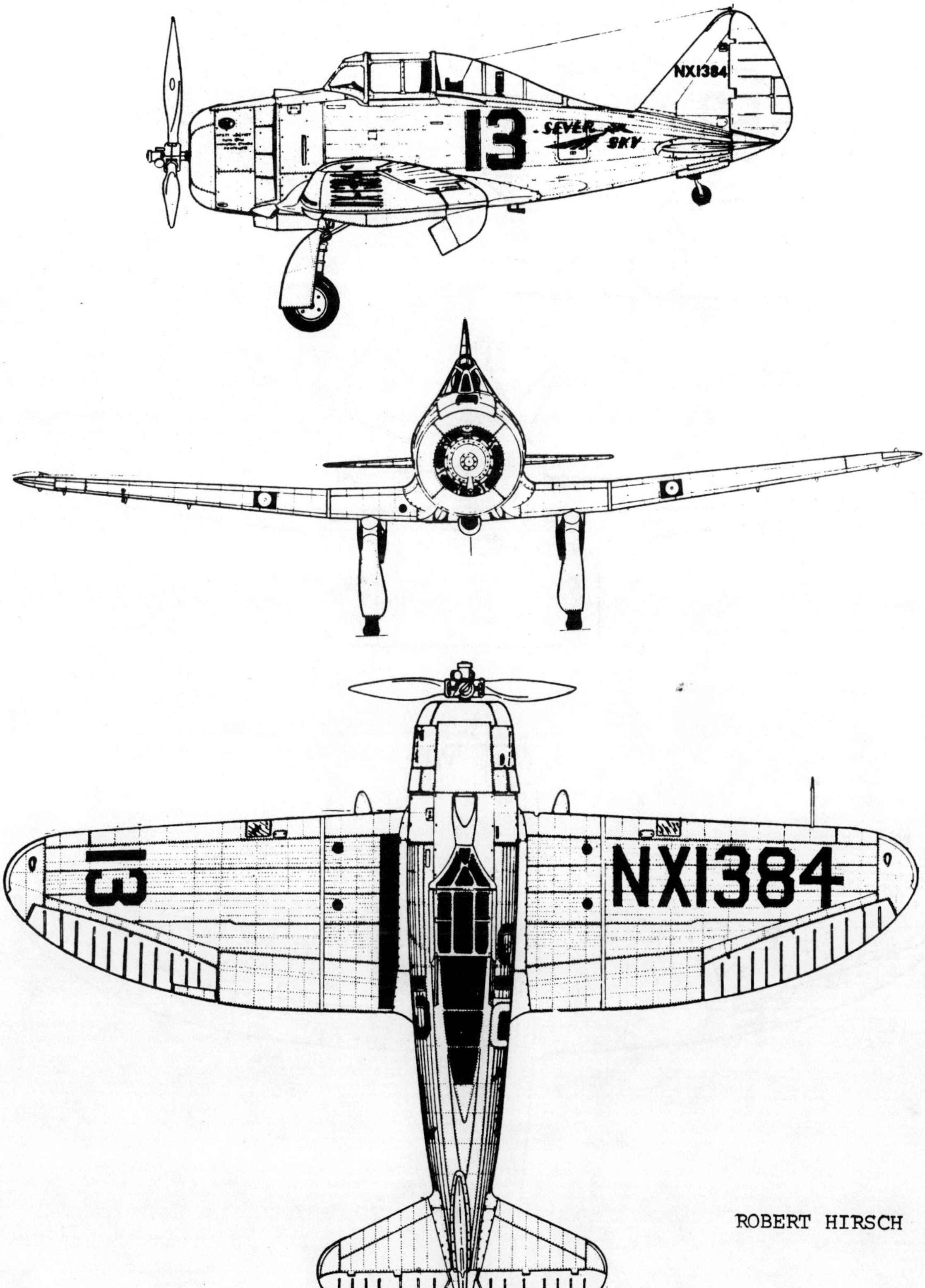

ROBERT HIRSCH

Seversky AP-7
Jacqueline Cochran's 1938 Bendix Trophy Winner

SEVERSKY EP1-106 SWEDEN

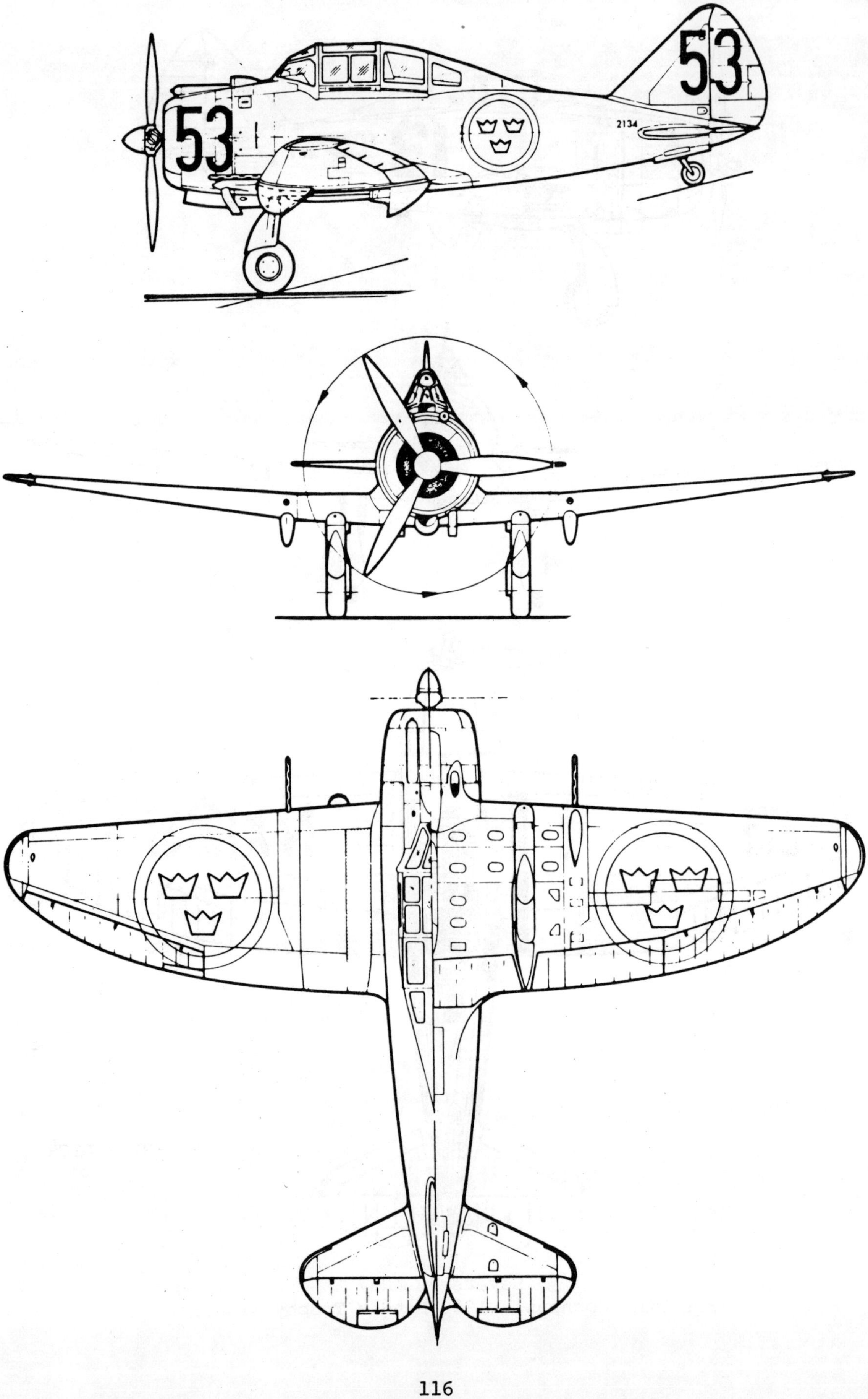

AT-I2 GUARDSMAN

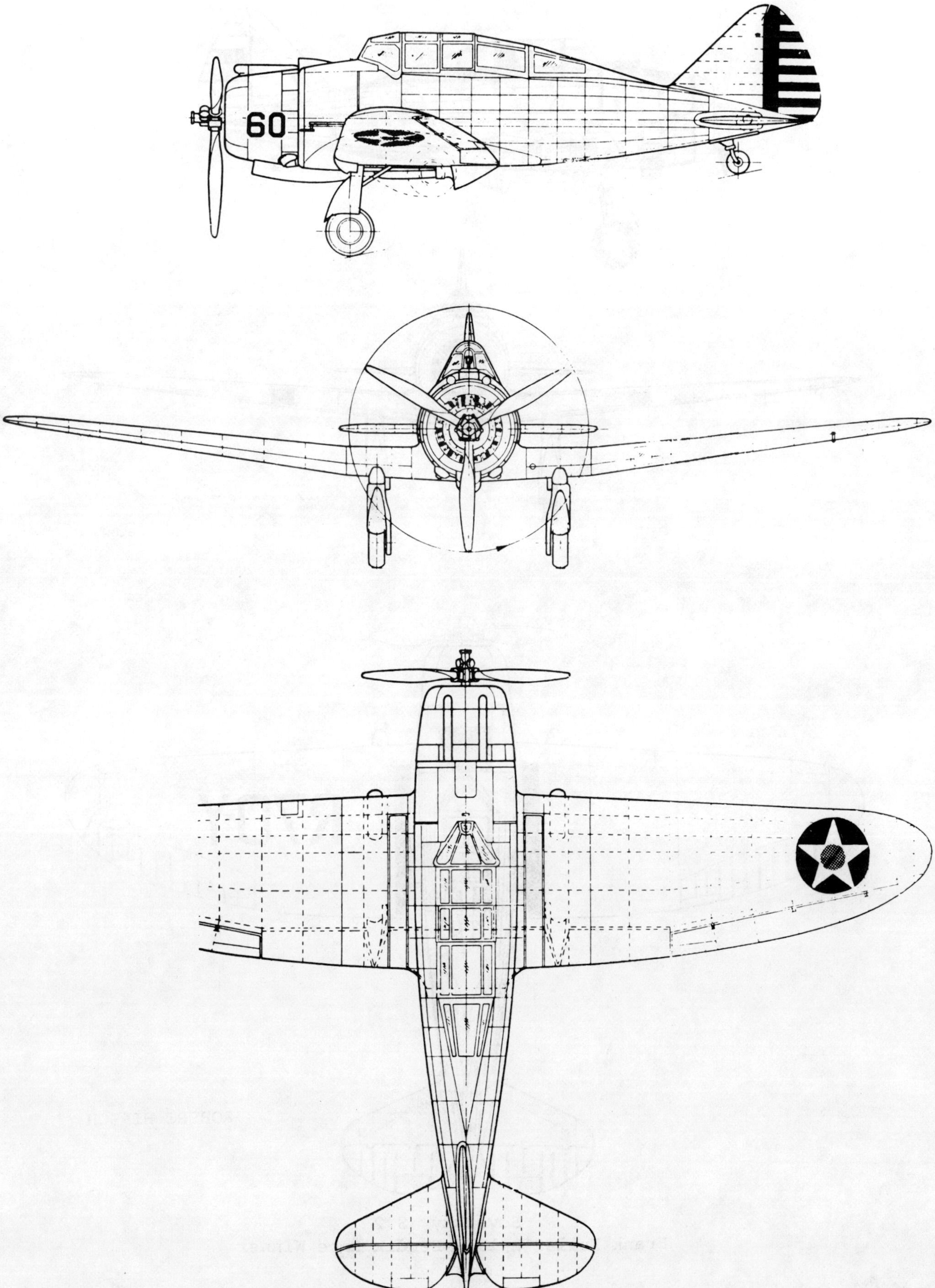

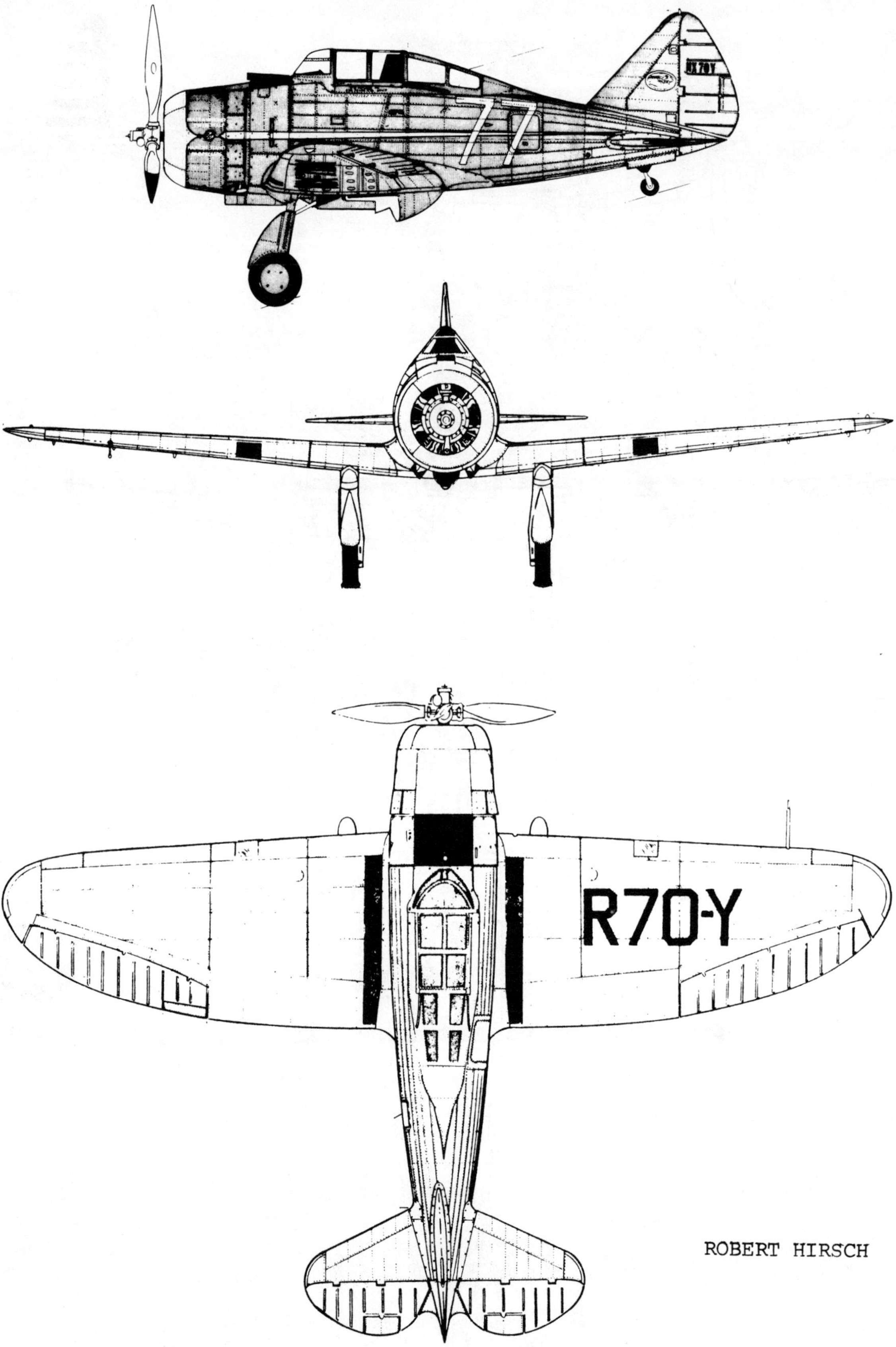

ROBERT HIRSCH

Seversky S 2
Frank Fuller's 1939 Bendix Race Winner

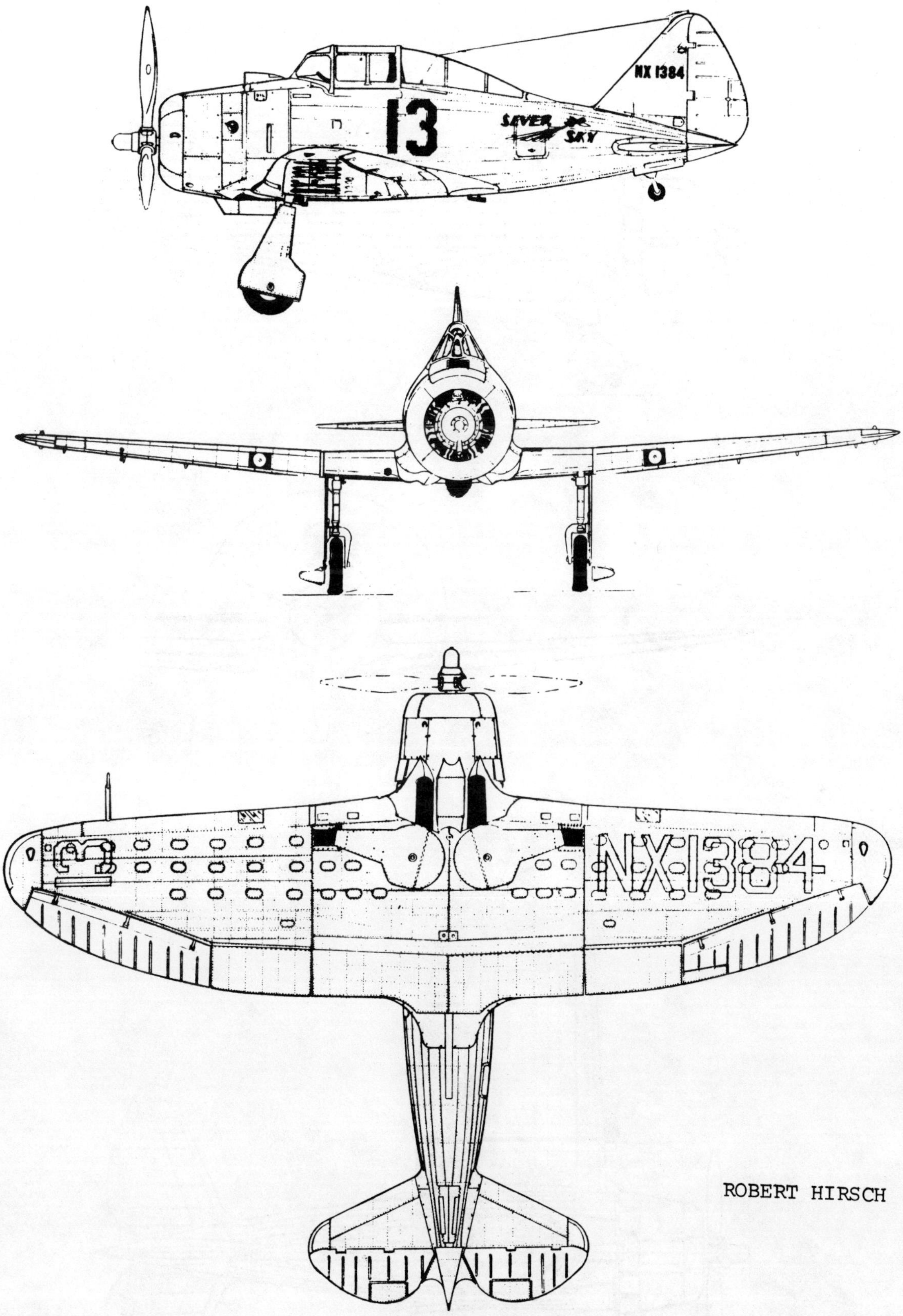

ROBERT HIRSCH

Seversky AP-7
Jacqueline Cochran's 1939 Bendix Race Entry

XP-4I EXPERIMENTAL PURSUIT

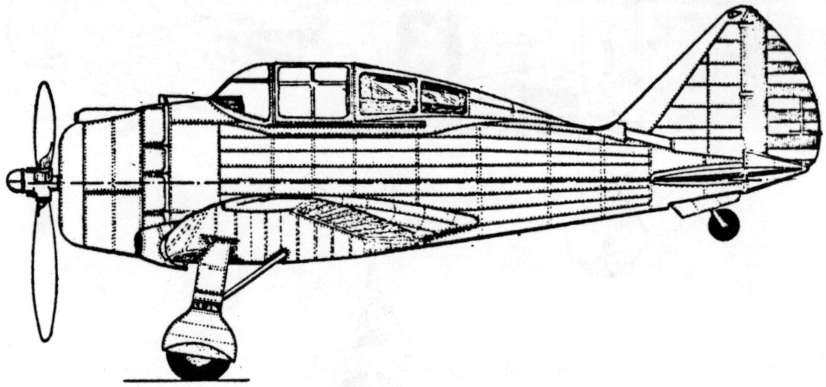

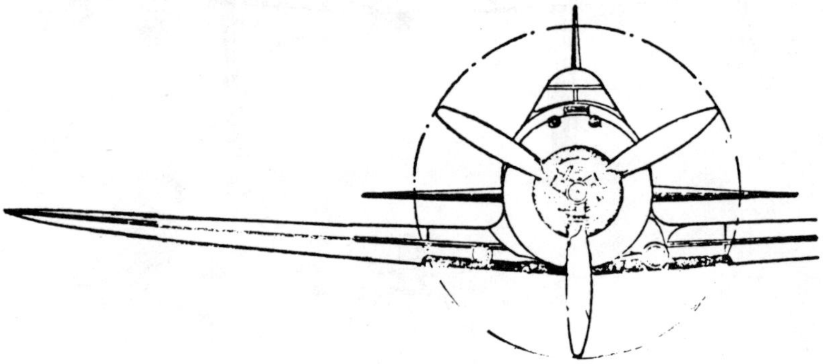

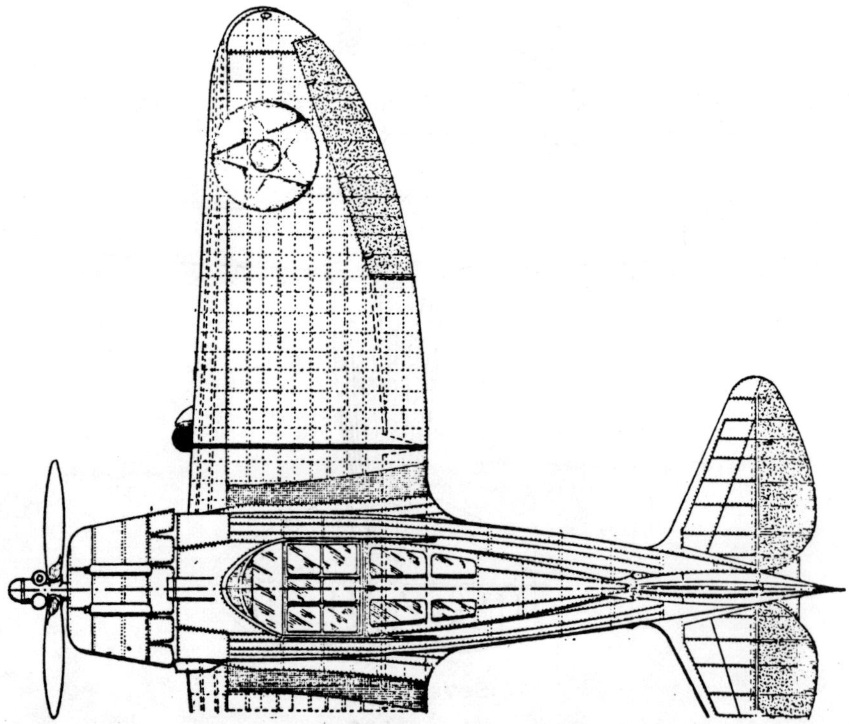

P-43A "LANCER"

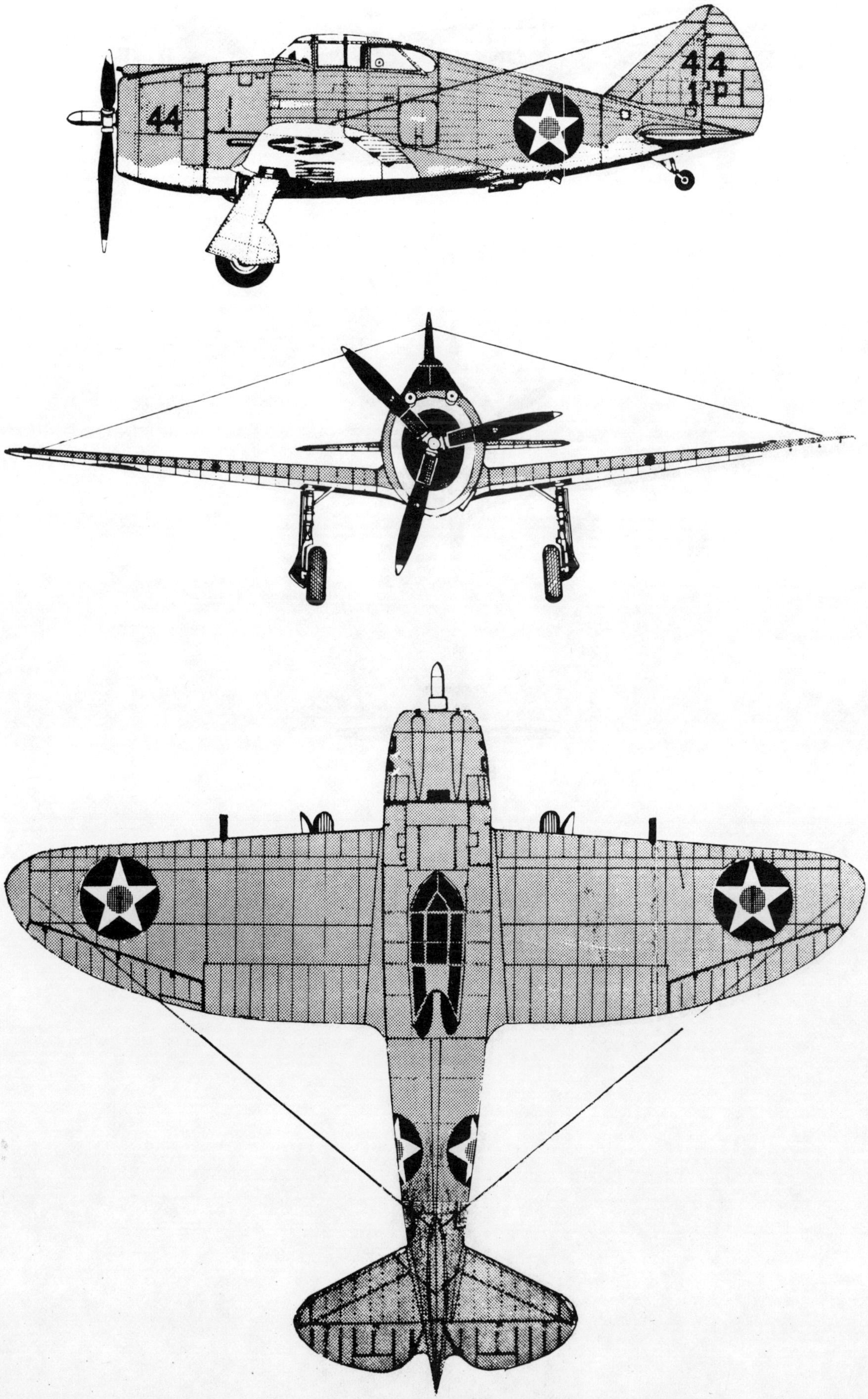

XP-44 "WARRIOR"

YP-47B "THUNDERBOLT"

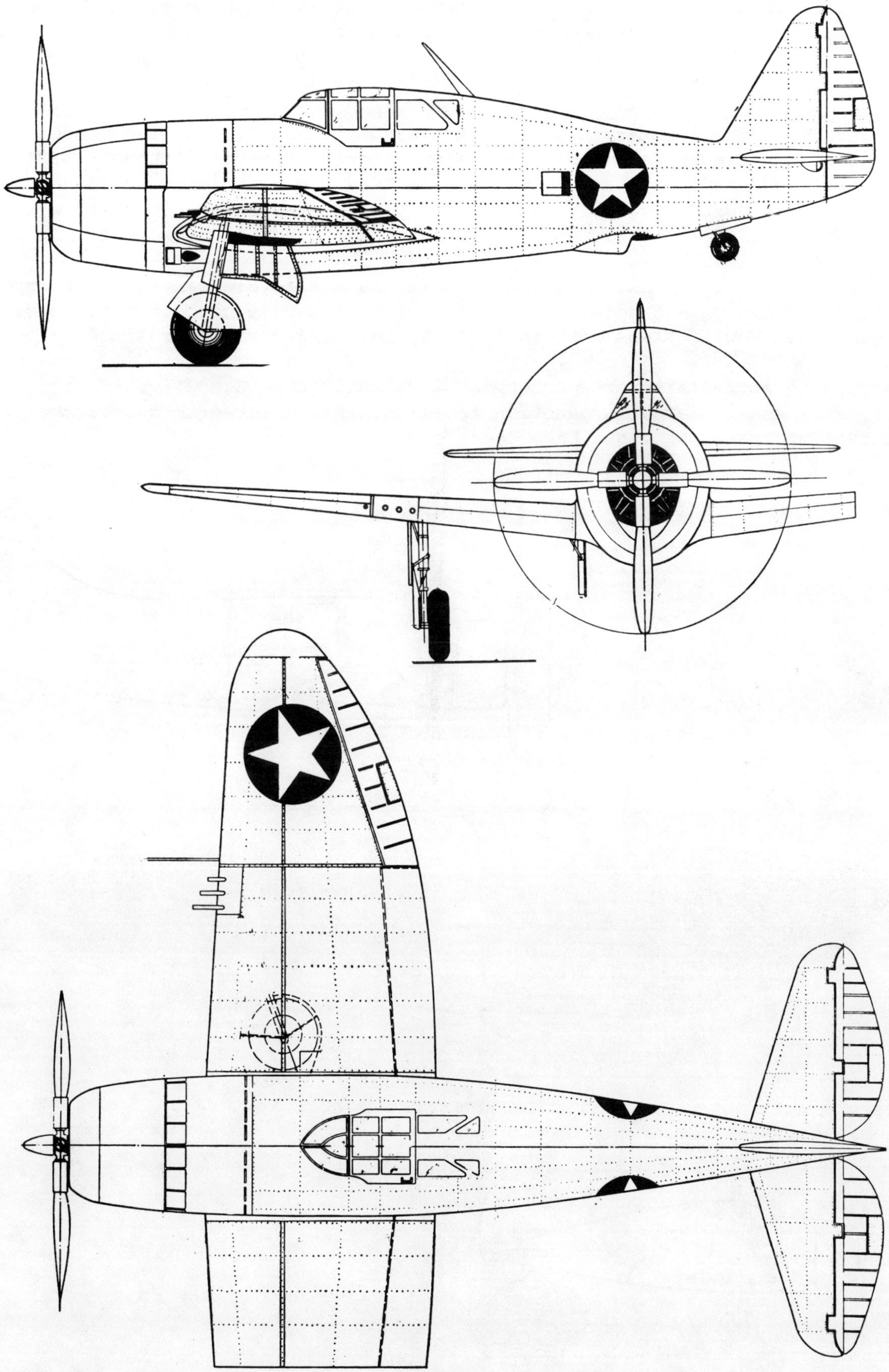

SEVERSKY EXECUTIVE TRANSPORT

During the mid thirties Seversky saw the need for a fast,economical executive transport for the business man. His first design had a wing span of 45 feet, and a length of 32 feet. Due to the pressure of military contracts no prototype model was built, but in mid 1937 he again proposed a new executive transport design utilizing the basic wing of the P-35 and landing gear, and tail wheel.
This aircraft was a luxary type closed cabin, all metal, low wing monoplane. The standard "Executive" model was designed to take Pratt-Whitney R-1830 engine of 1200 h.p., but any radial engine of 450 h.p. or more could be fitted. This gave the "Executive" a rated speed of 340 m.p.h. at 18,000 feet, which at the time was 120 m.p.h. faster than current scheduled air transports. Range was almost 2800 miles.
Late in 1938 Seversky incorporated a retractable tricycle landing gear into the "Executive" design and added large spinner to the nose cowl to provide better streamling. Fowler flaps were included in the design to reduce landing speed.

Although no actual aircraft prototypes were built the "Executive" designs did have merit and if larger production facilities had existed this Seversky "Executive" transport might have been constructed.

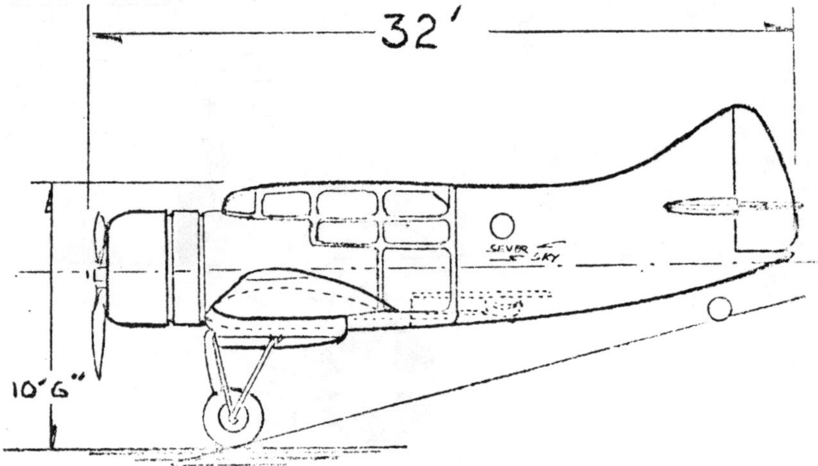

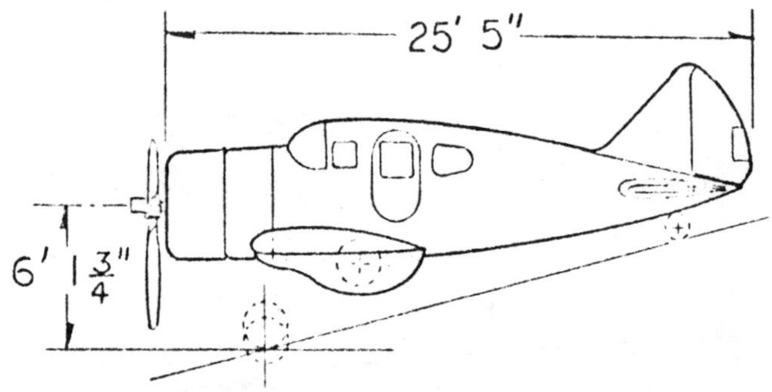

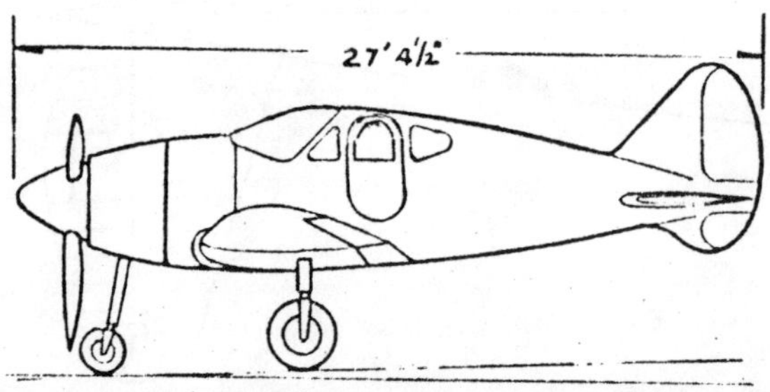

P-35 No.41 of 27th Pursuit Sq.,1st Pursuit Group. Most P-35's were based at Selfridge Field, Michigan. The P-35 was a beautifully constructed aircraft. It had a good rate of speed, long range and was sturdy built.

Original Seversky aircraft factory on Long Lsland, New York. This plant was originally the site of American Aircraft. Seversky shared this field with Grumman Aircraft and Ranger Engine Company. Grumman later moved to Bethpage, Long Island, New York.

IX . COLOR AND MARKINGS

The first Seversky built aircraft, the SEV-3 Amphibian of 1933, was painted gold all over with natural aluminum finish window and canopy frames. The majority of Seversky aircraft built after the SEV-3 were left in natural aluminum metal and were buffed to a high sheen.

The 30 BT-8 Basic Trainers delivered to the Army Air Corps were painted in standard Army training colors of the period with blue fuselage and chrome yellow wings and tail. The national markings were painted above and below each wing tip, but not on the fuselage. On the rudder the single blue stripe is painted vertically and red and white stripes are painted horizontally. Aircraft numbers were painted on the side of the fuselage and cowl in white.

U.S. Army Air Corps Markings: During the summer of 1937 a new aircraft identification marking system was introduced by the Air Corps. G.H.Q. Air Force established a marking guide to identify each aircraft by assigning two distinctive letters and a number to form an identification group. The two letters identify the purpose and organization. The number identifies the aircraft itself. Examples are:

PA-69 -No. 69 aircraft of the 1st Pursuit Group
"P" for Pursuit
"A" for 1st Pursuit Group
AC-17 -No. 17 aircraft of 3rd Attack Group
BG-3 -No. 3 aircraft of 7th Bombardment

For tactical squadrons not a part of a group, the group letter was omitted, and a certain number is assigned to the squadron for identification.

The first Army Air Corps squadron to receive the P-35 was the 94th Pursuit Squadron, 1st Pursuit Group, based at Selfridge Field, Michigan. P-35 aircraft of the 94th Squadron had one third of the nose cowl ring painted red. The squadron leader had two red bands painted around the fuselage. Each band was five inches wide and five inches apart and the bands were painted just aft of the pilots canopy. The "B" flight and "C" flight squadron leaders carried just a single red five inch fuselage stripe.

Identification designators of letters and numbers were used by all Army Air Corps squadrons. 94th Pursuit Squadron P-35 aircraft were assigned PA-70 through PA-99 inclusive. "P" indicated pursuit mission, and the "A" being the initial letter in the alphabet, stands for one, so that the whole may be interpreted as "1st Pursuit." The 94th was a component of the 1st Pursuit Group at Selfridge Field along with the 17th and 27th Pursuit Squadrons.

The complete designator number, say PA-70, is painted both above and below the left wing. Markings above are about two thirds the diameter of the national insignia in height, and are placed even with the top of the circle about midway between the insignia and the fuselage.

On markings underneath the wing, the numbers are placed in relation to the word "Army," as follows: Leading edge, PA-70, Army, trailing edge. Under the right wing the letters "U.S.," appear, so that the whole name "U.S. Army," is seen beneath the wings.

"PA-70" is also painted on both sides of the fin, in fraction style, with PA above the number 70.

All planes in the squadron carry the insignia on the fuselage, which is the head of an Indian in the act of "war whooping." The feathers of the headdress are white, with blue tips.

The other two Pursuit Squadrons of the 1st Pursuit Group were assigned the following aircraft numbers:

17th Pursuit Squadron: Numbers 11 through 39.
27th Pursuit Squadron: Numbers 40 through 69.

The 17th Pursuit Squadron had white cowl rings and squadron leader fuselage stripes. The "Snow Owl" insignia was painted on the sides of the fuselage. The 27th Pursuit Squadron had yellow cowl rings and squadron leader fuselage stripes. The diving hawk of the 27th Pursuit Squadron was painted on the sides of the fuselage. Designator markings were black on all three squadron aircraft.

War Game Camouflage: During the 1940 Army war games held in North Carolina P-35 aircraft of the 27th and 94th Pursuit Squadrons were painted in a temporary water base camouflage scheme of multi-colors of different hues of sand, olive green, and olive drab paints. The painting appeared to have been applied with brushes and patterns varied.

Seversky Racing Aircraft: When the Seversky SEV-3 amphibian was entered in the 1935 Thompson Trophy Race it was still painted in its all gold finish. It had black registration number on wings and tail.

When the Seversky IXP was completed and rolled out of the Long Island factory, the aircraft featured a straight wing. At this time the wings and tail of the aircraft were painted a pale blue. The fuselage was painted silver.

Later, when wing dihedral was added the aircraft was painted silver all over with black registration and racing number 63. It remained this all silver color until 1937-38 when the wings and tail were painted a gloss white. The fuselage retained its all silver color.

Jacqueline Cochran's Seversky AP-7 when it appeared at the 1938 Bendix Trophy Race was natural metal finish all over with a green number 13 outlined in white. The registration number, company emblem, and Pratt-Whitney lego were in black. For the 1939 Bendix the aircraft still retained its natural buff metal surface. The green racing number 13 was still retained however it was smaller in size than the 1938 number. It was outlined in white. The logo on the cowl was replaced by Pratt-Whitney emblems. Registration number and company emblems were black.

For the 1937 Bendix Race, Frank Fuller's Seversky S-2 racer was left in natural metal finish. It bore race number 23 in black on the fuselage. All registration and other details were in black. For the 1938 Bendix Race, Frank Fuller had his race number changed to 77. This was painted in black on the fuselage. The aircraft was left natural finish all over, however the nose cowl front was painted green as was the stripe running down each side of the fuselage. Registration number and logo were black.

As president of one of America's largest painting companies, Frank Fuller painted his racer an entire new color scheme for the 1939 Bendix Race. His Seversky S-2 Racer was painted entirely in the new green metallic paint which had just been marketed. The racer also featured a yellow nose cowl fuselage stripe and race number. Registration was black. The propeller was polished metal with the rear upper half of the blades painted flat black to reduce glare reflection in the pilots eyes.

EXPORT MODELS: Almost all of the Seversky export model aircraft were delivered in natural metal finish.

Japan: The Seversky 2PA-B3 aircraft exported to Japan in 1938 were delivered in natural metal finish. As these aircraft were destined for service in China they were painted in Japanese camouflage patterns of green and brown on the top side and a grey bottom.

Sweden: The Seversky EP-106 and 2PA-204A aircraft were delivered to Sweden in natural metal finish. The Swedish Air Force later painted the aircraft in dark green on the upper surface and a pale blue on the bottom surface.

World War II

Philippine Islands: P-35A's in the Philippines were delivered in natural metal finish complete with Swedish Air Force markings. The United States insignia was applied over the Swedish insignia and tail markings were applied to the rudder. All P-26A and P-35A pursuit aircraft in the Philippine Islands were given the designator "4MP." This marking was carried on the same locations as "PA " markings were carried on P-35 pursuit aircraft as previously described.

During the summer of 1941, a program was initiated to camouflage the P-35A's of the 21st Pursuit Group in a green and brown camouflage. The only paint stocks available in the islands was a water based paint, which worked perfectly well until one flew into a weather front and all the paint washed away. The majority of P-35A aircraft had been camouflaged by the time the Japanese struck the Philippines, however as one Zero pilot once commented after hitting Clark Field, "The sun's reflections off the silver wings of the aircraft on the ground could be seen from over 50 miles away."

Some of these aircraft were Boeing B-17C and B-17D bombers as well as P-35A's.

United States: After the outbreak of the war, stateside Seversky P-35 P-35A, and P-43 aircraft were painted olive drab color top side and grey on the bottom. Propellers were painted flat black as were the serial numbers on the rudder. The few remaining P-35's were replaced as new fighters emerged from the factories. The P-35A's were sent to Ecuador. The few P-43's remaining went to fighter training schools and ended the war in olive drab and grey camouflage.

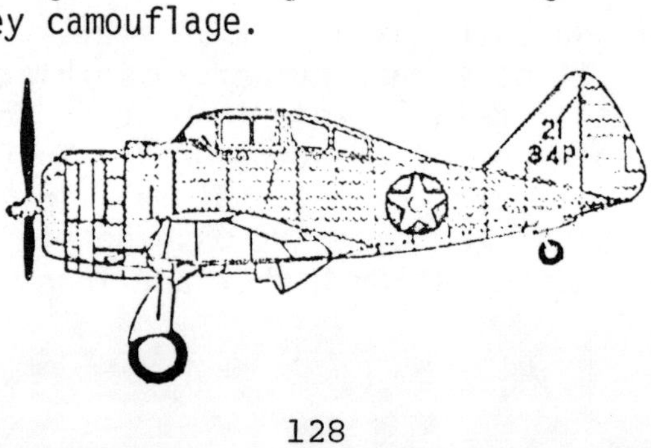

X MODEL EVOLUTION AND PRODUCTION HISTORY

Early Seversky model designations used the prefix "SEV" and a number and letter designation to identify seating capacity, and sometimes customer and design missions. The various types of aircraft manufactured were not listed in sequence development.

The first Seversky aircraft built, the SEV-3 of 1933 was a three place amphibian. Later, when the floats were removed and a fixed landing gear was fitted it became the SEV-3XAR (3 Place, EXperimental Army) and was demonstrated to the Army Air Corps. According to Maj. Seversky the first number signified the number of persons the aircraft carried, and the letters that followed signified the mission. Example: The SEV-2XP was a two place, experimental pursuit for the Army.

Later designations incorporated the engine as well. Example: The first aircraft export contract received by Seversky was from Columbia. These aircraft were designated SEV-3M-WW (3M stood for 3 place Military, and "WW" stood for Wright Whirlwind Engine.

The reader must bear in mind that most major components of Seversky designed aircraft were interchangeable with other Seversky aircraft. Parts from one aircraft might appear on another Seversky aircraft months or years later to meet the company's needs at that particular time.

Aircraft built for special customers like Jimmy Doolittle bore the designation SEV-DS or "Doolittle Special." The Seversky built for Frank Fuller was designated SEV-S 2 or "SEversky Executive."

The Seversky 2PA was a two place aircraft for export. Letters following 2PA models signified type of aircraft or mission.

It was not until 1937 that a new designation system was introduced. Designations were now formed in logical sequence. Thus AP-1, and AP-2 were Army Pursuit designs with radial engines, AP-3 was an Army Pursuit design with a liquid cooled engine. It was never built. The AP-4 was Army Pursuit model four, etc., etc.

Some confusion existed but the reader must bear in mind the year in which the aircraft first appeared. The EP-1 or "European Pursuit 1st Model" was the export model for the Seversky AP series. The 1st letter signified customer, the 2nd letter signified mission, and the sequence number signified the model.

PRODUCTION HISTORY

YEAR	c/n	MODEL	REGISTRATION SER./NUMBER	REMARKS
1933	1	SEV-3	X-1206	3 Place Amphibian, Wright J-6 420 h.p. Engine.
1934		SEV-3XAR	X-2106	Modified to Land Plane with fixed spatted gear, Wright R-975 engine, 350 h.p. Won BT-8 contract.
1935		SEV-3M	X-2106	Mod.Back To Amphibian,R-1820,710 h.p. Set Speed Record 9-15-35 230.4 mph. Sold To Mexico Army Man.
1935	2	SEV-2XP	X-18Y	2 Place Exp. Pursuit, Wright R-1820 800 h.p. Damaged in transit to Wright Field 6/36.
1935		SEV-1XP	X-18Y	2XP Returned To Factory & Became 1XP. Returned to Wright Fld.Competition,Raced by Sinclair,Scrapped.
1935	3 4 5	SEV-3M-WW " "	NX-15391 NX-15689 NX-15928	1st Foreign Order for Columbia,3 Place Amphibian. U.S. Reg.Shows c/n's as 1,2,&3 but were actually c/n 3,4,&5 aircraft by factory.
1936	6	SEV-X-BT	X-189M	Exp.2 Place Trainer,R-1340,550 h.p. Lost out to BT-9 in 1936 Competition.Later mod.to 1st 2PA. Toured S.Amer.1938.Broken up for spare parts.
1936	7-36	BT-8	35-247 thru -276	30 Mfg.for Air Corps 1936. P&W R-985 engine 400 h.p
1936	37-39	SEV-3M-WW	- -	2nd Order,3 Place Amphibian for Columbia.
1936	40	AP-2	R-1250	Basic 1XP,Crash landed at Floyd Bennet Fld.N.Y by Maj.Sev.just prior to 1937 Cleveland Races.
1937	41	NF-1	NX-1254	Entered U.S. Navy Fighter Competition 1937,R-1820.
1937	42	SEV-DS	NR-1291	Especially Built for Jimmy Doolittle,Shell Oil Co.
1937	43	SEV-S2	NR-70Y	Built for Frank Fuller, Basic AP-1 Design. Won 1937 & 1939 Bendix Air Races, P & W R-1830 engine.
1937	44	AP-1	NR-1390	P-35 Demo. Aircraft,Bulged Windshield,R-1830, Later Fitted With Tight Cowl & Large Spinner.
1937	45-121	P-35	36-354 thru 430	1st Air Corps production order.R-1830-9. 1st Deliveries to 1st Pursuit Grp.Selfridge Field.
1937	121	XP-41	36-430	Last P-35 Delivered As XP-41, Hi Alt. Ftr.R-1830.
1937	122-141	SEV-2PA-B3	-	2nd.Export Sale,Wright R-1820,20 To Japan,Became A8V-1,Served in China,Few To Asahi Newspaper.
1937	-	AP-3	-	Designation of Ftr. With Allison Engine.Never Mfg.
1938	142	SEV-2PA-A	NX-1304	3rd.Export Sale, 2 Place Amphib.Ftr. for Russia Mfg. Rights Sold For Reported 1 Million Dollars.
1938	143	SEV-2PA-L	R-189M	2 Place "Convoy Ftr." Sold Russia, saw service in Spanish Civil War on Republican side.
1938	144	SEV-2PA-L	-	Also Believed Sold to Russia for service in Spain.

YEAR	c/n	MODEL	REGISTRATION SER./NUMBER	REMARKS
1938	145	AP-7	NX-1384	Army Pursuit No.7,Fully Retractable Gear, J.Cochran Won 1938 Bendix Race. Aircraft Destroyed In Hangar Fire, Tampa, Florida.
1938	146	SEV-2PA-BX	NX-2586	2 Place "Convoy Ftr." Demonstrated Europe But In Crash in England April 1,1939.
1938	147	SEV-1-68	NX-2587	Similiar To AP-7,Demo.In Europe & Led To Large Swedish Order. Aircraft Sold Dom.Republic 1939
1939		AP-4	NX-2597	Entered Wright Fld.Competition March 1939. Inward Retracting Gear,Flush Revited,Turbo S.C. Later fitted with tight cowl & Big Spinner.

When Maj. Alexander P.de Seversky was relieved as President of company on Oct.12, 1939 the new company now known as Republic Aviation changed c/n designation system.

YEAR	c/n	MODEL	REGISTRATION SER./NUMBER	REMARKS
1939		AP-9	NX-2598	Entered Wright Fld. Design Competition Feb.-April 1939,J.Cochran Set 2000 km. Speed Record of 322 mph. Sold to Dominian Republic 1940.
1939		YP-43	39-704 thru 39-716	Company Project No. 79, Hi Alt. Ftr. R-1830-35 1200 h.p., Hit 351 mph. at 20,000 Ft.
1940		EP 1-106 P-35A	41-17434 thru 41-17493	Company Project No.80, 60 Seized By U.S. Gov't Oct.1940 from Sweden.45 Sent Philippine Is. For Defence.
1940	483-01 483-02	2PA-204A	-	Two 2PA Aircraft Delivered to Sweden From 52 Originally Ordered On Contract.
1940	483-03 thru 483-52	2PA-204A AT-12	41-17494 thru 41-17543	50 Seized By U.S. Gov't.Company Proj. No.83 Designated AT-12 "Guardsman" R-1830-45
1940		P-43	41-6668 thru 41-6721	Republic "Lancer" ,Company Proj. No. 88, Hi-Alt. Ftr. with Turbo Super Charger
1940		P-43A	40-2891 thru 40-2970	Original XP-44 Order Cancelled, 80 P-43A's Substituted. Company Proj. No 90, R-1830-49.
1941		P-43A-1	41-31448 thru 41-31572	Final Production Order of "Lancer." Company Proj. No. 92. Some Mod. to Photo-Reccon, Majority Sent Chinese Air Force 1942.
1941		XP-47B	40-3051	Allison Powered XP-47 Ftr. Was Rejected. Light Wt. XP-47A Ftr. Also Rejected. Company Proj. No. XP-47B 1st Flew 5-6-41, R-2800-21 2000 h.p., Max. Speed 412 mph. Prototype of Famous W.W.II "THUNDERBOLT" Fighter. Some 15,579 mfg. during 1941-1945.

SPECIFICATIONS

Year	No. Mfg.	Model	Reg.No.	Engine	H.P.	Max.Speed	Gross Wt.	Span	Length	Ht. Ht.	Notes:
1938	1	SEV-3	X-2106	J-6/R-975	220/350	190mph@ S.L.	3,250 lbs.	36'o"	25'8"	10'	Held Speed Record
	(1)	SEV-3 XAR	"	R-975	350	200 " " "		36'o"			Fixed L. Gear
1935	(1)	SEV-3M		R-1820	715	230mph	5,000 lbs.	36'o"	28'10"	10'	Speed Record 1935
1934	1	SEV-2XP	X-18Y X-18Y	R=1820	850	240mph		36'0"	24'1"	9'9"	2 Place Pursuit
1935	(1)	SEV-1XP	X-18Y	R-1820	850			36'0"	24'1"	9'9"	Speed Record A/C
1936	30	BT-8	-	R-985-11	400	175mph	4,050 lbs.	36'0"	24'4"	8'10"	1st Low Wing Trainer
1936	1	SEV-XBT	X189M	R-1340	450	180mph	4,500 lbs.	36'/41'	30'5"		
1936	77	P-35	-	R-1830-9	950	281mph@ 10K	5,600 lbs.	36'0"	25'2"	9'9.5"	1st Ftr. Order
1937	1	NF-1	X-1254	R-1820	950		5,000 lbs.	36'0"	25'2"	"	Tested By U.S.N.
	6	SEV-3M-WW	_	R-1820	715		5,000 lbs.	36'0"	28'10"	10'	To Columbia
	60	P-35A		R-1830-45	1000	290mph@ 12K	6,720 lbs.	36'0"	26'10"	9'9"	Seized By U.S.Gov't
	60	EP1-106		R-1830	"						To Sweden
1938	2	SEV-2PA-L	NR 189M	R-1820	850	250mph	5,900 lbs.	36'0"	30'5"	10'9"	To Russia
1938	1	SEV-2PA- A		R-1820	850	230mph		36'0"	30'5"	11'3"	To Russia
1938	20	SEV-2PA-B3		R-1820	850	250mph	5,500 lbs.	36'0"	25'3"	9'9"	20 To Japan
1938	1	SEV-2PA-BX		R-1830	1000	300mph +	5,000 lbs.	36'0"	26'10"	9'9"	Crashed in England
1939	50	SEV-2PA-204A (AT-12)		R-1830-45 R-1830-45	1000 "	295mph	7,500 lbs.	41'0"	27'8"	9'10"	Seized By U.S. Gov
1938	(1)	XP=41		R-1830-19	1200	320mph	6,500 lbs	36'0"	27'0"	10'6"	Hi-Alt. Pursuit
1939	268	YP-43,P-43 P-43A		R-1830	1200	350mph	7,155 lbs.	36'0"	28'6"	10'2"	Production Model
1940	-0-	XP-44		R-2180/ R-2800	1400/ 1850	-	-	36'0"	29'3"	14'2"	Exp. AP-4J/AP-4L Design Only
1941	1	YP-47B		R-2800-21	2000	400mph +	12,000 lbs.	40'9"	35'0"	14'2"	1st of Mighty "Thunderbolts"

XII MILITARY SERIAL NUMBERS

No. Mfg.	Model	Serial Number	Contractor Number
30	BT-8	35-247 - 35-276	c/n 7 - 36
77	P-35	36-354 - 36-430	c/n 45 - 121
60	P-35A	41-17434 - 41-17493	-
(1)	XP-41	(36-430)	c/n 121
50	AT-12	41-17494 - 41-17543	c/n 483-03 - 483-52
13	YP-43	39-704 - 39-716	
54	P-43	41-6668 - 41-6721	
80	P-43A	40-2891 - 40-2970	
121	P-43A-1	41-31448 - 41-31572	
1	XP-47B	40-3051	

IX TEST PILOT REPORT Model 2PA-204A

By Capt. Don Lykins

The flight of this aircraft was made to give an accurate evaluation of this famous aircraft's design and handling characteristics for a fighter of the 1936-1940 period of aviation development.

The aircraft is a two place, long range, export version of the Seversky Pursuit with the basic differences being that it has a rear cockpit and extended wings to accommodate outboard fuel tanks.

Sitting in the cockpit, one is impressed with both the size and the orderliness of design that was extremely unusual for aircraft of this era. When sitting in the cockpit with the canopy open, the pilot is very comfortable. It might be well to mention here that the Seversky P-35 was the first production built pursuit aircraft for the Army Air Corps to have an enclosed cockpit. The canopy when closed provided excellent streamlining for the aircraft, however the pilot must lower the seat considerably. The forward visibility from the pilots cockpit as compared to say the early Air Corps Boeing P-12E pursuit, is quite good with the only major obstruction being the carburetor air scoop which restricts the straight over-the-cowling visibility. The cockpit is an exceptionally clean layout with the throttle, propeller, and mixture control all close at hand and grouped comfortably together on the pilot's left. Just below the throttle quadrant is the power selection for hand or electric operation of the flaps and gear. It is a

KEEP 'EM FLYING

simple and very advanced system that provides for both normal power and emergency hand crank capability of the system. Located just slightly below the flap and gear controls are the gear and flap indicators which are cable driven and very easy to read. The electrical panel at the pilot's left is convenient and has the boost pump, amp meter, starter, battery switch, etc. The fuel selector is located on the pilot's lower left and is a very simple straight-forward fuel system with five positions: right and left aux, the main tank, the reserve tank, and the off position. To the right front of the pilot and located at the base of the instrument panel is a crank handle for manual cowl flaps which are very simple to operate.

Starting the Pratt-Whitney R-1830 "Wasp" engine is quite simple by just turning the battery switch on, switching fuel on to the fullest tank, turning on the boost pump, putting the mixture to rich, allowing the fuel pressure to rise, then shutting off the boost pump. At this time, you engage the starter, allow nine blades for cylinder scavaging, turn the magneto switch on and the engine fires quite quickly.

Taxiing the Model 2PA as compared to other fighter type aircraft of this era is excellent with good brakes and a fully steerable tailwheel. The visibility is fair with the size of the cowling being somewhat deceptive as to how much area a pilot can see. Visibility however, is quite good overall. Like most all tail dragger pursuits, "S" turns must be made on the ground while taxiing.

A normal takeoff was made. The aircraft handles extremely well, with very positive directional control on takeoff and little or no tendency to wander on the runway. The initial climb was at 2300 r.p.m., 35 inches,

150 m.p.h. indicated and we had a steady 1500 to 2000 feet a minute rate of climb. During the takeoff and climb period, the wisdom of the cockpit designer in placing everything close at hand makes this a very uncomplicated, undistracting aircraft to fly. The Pratt-Whitney R-1830 is very smooth and allows exceptionally small amounts of vibrations to come back through the aircraft itself. The exhaust stacks extend vertically below the engine, making the aircraft exceptionally quiet in the cockpit for this type of aircraft.

I did climbing turns with the aircraft, 30 and 45 degrees of bank and the aircraft is quite comfortable in lateral control. It is rather heavy on the ailerons. It is very sensitive on the stick and pitch forces and the rudder is very sensitive and responsive.

I set up initially at 13,000 feet and set a slow cruise power of 1725 r.p.m., 25 inches and speed settled down to indicating 170 m.p.h. The nose sits rather low on the horizon giving the pilot somewhat of a false sense of descending in level flight. However, this illusion is overcome quickly and the aircraft trims hands off and is very stable in cruising flight.

STALL CHARACTERISTICS

I conducted a series of stalls with the aircraft, both power on and power off, clean and dirty, and in level and turning flight. The following is the results of those examinations.

With the power off, straight ahead, gear and flaps retracted, you get a good warning with stick and airframe buffeting beginning at 100 to 105 m.p.h. This is followed by a tip stall with a light roll either right or left at about 90 m.p.h. with a complete stall starting at the tip at 80 m.p.h. Recovery is quite easy. Just relax the back pressure and add a small amount of power. I attempted a more rapid recovery to see the secondary stall characteristics clean, and found that a rather sharp tip stall occurs at 100 m.p.h. with the accompanying roll if you pull too quickly during the recovery.

The next stall was a power on stall with approximately 1700 r.p.m. and 25 inches of power in a 25 degree bank. The first one was to the left. Putting the aircraft in fair trim during the turn at 100 m.p.h. indicated air speed, there is no advanced air frame buffeting and no stick buffeting as was experienced in the straight ahead stall. The aircraft brakes very quickly over the top to the right, rolling with an easy recovery by relaxing the back stick pressure, lowering the nose slightly and leaving the power in the same position as the entry.

In the same configuration, a turning stall now to the right, the aircraft in a slight right skid stalls with no warning and will spin quickly into the turn if unchecked. Recovery is easy. Relax the back pressure, allow the air speed to increase smoothly, return the nose to the horizon.

The next stall was straight ahead with the flaps and gear in the down position. The power was 1700 r.p.m. Manifold pressure at 15 inches. At 75 m.p.h. indicated air speed, the wing stalls again from the tip in. There is no advanced air frame or stick buffeting as in the case of the clean stall. The only stall warning occurs when ailerons begin to buffet and the aircraft rolls right or left depending on which tip stalls first.

All in all, the airplane displays quick braking stall tendencies but no indication or tendency to resist a recovery. The stalling characteristics are a somewhat classic example of aircraft design during this period.

ROLLS

I did a series of rolls with the aircraft, and experienced the following conditions.

The ailerons are very long and to roll to either right or left takes a heavy control pressure. It is much heavier than other aircraft designed previous to this era such as the P-12 and P-26 pursuit type aircraft which are extremely light in aileron control. I found to maintain a brisk roll rate was almost a two handed operation above 170 m.p.h. with the highest speed roll done at approximately 300 indicated. The aircraft would be classed as quite poor in roll due to the heavy pressures required on the stick. The aircraft rolls extremely well to the right and is rather uncomfortable in its rolls to the left. In constrast the rudders control and the stick forces are light, up through and including 300 m.p.h. indicated air speed.

GENERAL CHARACTERISTICS

I rolled the aircraft over and dove approximately a 40 degree dive angle with 2000 r.p.m and 30 inches of power at 13,000 feet and found that the aircraft is sufficiently dirty aerodynamically that it does not build up speed rapidly. This is more characteristic of the P-12, P-26 type aircraft than of other aircraft such as the P-40 design which has better acceleration in a dive. At approximately 300 m.p.h. indicated air speed, at 9000 feet, the noise level becomes quite high in the cockpit. The flap and gear operations of this aircraft affect the nose position somewhat as follows: The landing gear up, down, or in transit causes no pitch change on the aircraft. On flap extension after about a quarter flaps are extended as the flaps go down, the nose goes down. On retraction the flaps coming up cause the nose to come up only slightly but causes a very definite sink rate on the aircraft if not checked by stick forces by the pilot. This again is rather a normal flying characteristic of aircraft design in this era and was not particularly undesirable unless the pilot was making a night go around or an instrument go around and failed to check the sink rate established by the retracting flaps. Then this feature alone could be extremely dangerous to the unwary pilot.

APPROACH AND LANDING

During our stall test we established the stall speeds for our approach configuration at approximately 80 m.p.h. Adding 30 percent to that, we come up with an approach air speed of approximately 95 m.p.h. I entered downwind and abeam of the tower, extended the landing gear and held 150 m.p.h. Turning base leg, I slowed to 125 m.p.h. and extended approximately a quarter flap. Turning final, I slowed to 110 m.p.h. extending the balance of the flaps, and noticed the aircraft had an extremely high rate of descent capability with full flaps and power off. Crossing the end of the runway at 95 m.p.h. the aircraft was extremely responsive and roll control to counteract an approximately 15 m.p.h. crosswind that was blowing at the time. Rudder control was exceptionally good and visibility would be classed

as outstanding for an aircraft of this era. I touched down on the main landing gear, pushed the power button to retract the flaps, advanced the propeller to full r.p.m., and executed a touch and go with excellent control on all flight controls throughout.

On the next landing, using the same approach technique, I held 1000 feet above the airport until approximately ½ mile from the end of the runway. At that time I had full flaps, 110 m.p.h., and I had closed the throttle and the aircraft descended approximately 110 m.p.h. and was still obviously going to be high. I went into a full side slip to the left. The resulting rate of descent was probably 2000 to 2500 feet a minute and it was necessary to add power to get to the end of the runway. Again a touch and go landing was performed with outstanding control throughout.

On the third landing I positioned the aircraft approximately 1500 feet above the airport and approximately one mile out, came back on the power to approximately 20 inches, extended full flaps and conducted full side slips in both directions and again it was necessary to add power to get to the end of the runway. Control was excellent throughout the side slips with no tendencies for the aircraft to buffet or show any unusual signs of control difficulties. On touchdown I began braking while the aircraft was on the main gear to test the effectiveness of both the brakes and elevator response to braking on the main gear, and it was excellent for an aircraft of this type. With the large elevator control available it was possible to bring the aircraft almost to a stop and then gently lower the tail to the ground. I used approximately 1800 feet in landing and feel it could have comfortably stopped in approximately 1000.

In summary, the overall flying characteristics of this aircraft designed in the mid-1930's was in my estimation far superior to most any aircraft of its time. Its construction was all metal with the exception of the fabric covered control surfaces and by its systems and its design it had a built-in ease of maintenance that must have been applauded by most all mechanics. Its simplicity in systems was superior in design to most aircraft of this era, and all of its flight characteristics would be classed as good to gentle with the one exception of the high stick control forces required to rolling maneuvers with the aircraft.

Flying for annual August "Gathering Of Eagles" Warbird Air Show at Chino Airport, California. The last of its breed still flying.